THE TIME DIMENSION

THE TIME DIMENSION

An Interdisciplinary Guide

Compiled by T. K. Das

PRAEGER

New York
Westport, Connecticut
London

Library of Congress Cataloging-in-Publication Data

The Time dimension : an interdisciplinary guide / compiled by T. K. Das.
p. cm.
Includes bibliographical references.
ISBN 0-275-92681-8 (alk. paper)
1. Time—Bibliography. I. Das, T. K.
Z7128.T57T55 1990
[BD638]
016.115—dc20 90-31137

Library of Congress Catalog Card Number: 90-31137
ISBN: 0-275-92681-8

First published in 1990

Praeger Publishers, One Madison Avenue, New York, NY 10010
An imprint of Greenwood Publishing Group, Inc.

Printed in the United States of America

The paper used in this book complies with the Permanent Paper Standard issued by the National Information Standards Organization (Z39.48-1984).

10 9 8 7 6 5 4 3 2 1

To the memory of my father

Contents

Preface

The idea for this book of bibliographic items began with the search for published research on the subject of time that may have some relevance to my doctoral dissertation topic in the area of strategic management. The more I delved into the temporal literature, the more I became convinced of its many-splendored character. The need for an interdisciplinary compendium of temporal scholarship became readily obvious. The nature of research insights on the subject of time, perhaps in consonance with the nature of the phenomenon of time itself, was difficult to comprehend within the confines of any specific discipline. It seemed that scholars and others interested in the time dimension should find it eminently useful to have a compilation of some of the published works in different disciplines.

Such a comprehensive bibliography should hopefully lead the interested individuals to look up some of the literature far afield from their own particular province. Let me illustrate this with the subject I am most familiar with, namely, management and organizational studies (MOS). There is evidence of increasing interest in time related research in this area (see, for instance, some of the recent publications listed in Chapter 12). The inspiration for this temporal research -- conceptual, theoretical, and empirical -- has inevitably been the literature published in various other disciplines. However, there is presently no guide or compilation of the available publications concerning the temporal dimension that is in some degree relevant to research in MOS. A bibliographic guide would serve to meet the needs of the expanding network of scholars either actively pursuing temporal research at present or interested in exploring the subject. One of the reasons MOS scholars have thus far not been able to make any substantial headway is that the volume and diversity of the literature on the time dimension in different disciplines is very considerable. As can be surmised, a substantial portion of the published material is of little relevance to the temporal aspects of MOS. Much sifting is thus needed. Having been engaged in this field for several years now, I have attained an adequate degree of familiarity with the interdisciplinary literature to be able to select the more significant streams of research.

Given the vast amount of available literature and the relatively limited space in a book of this type, some critical choices had to be made about what to include and what to leave out (often very reluctantly). This is also the reason for not having abstracts, as that would have resulted in the elimination of substantial portions of relevant literature. A complete author index has however been provided.

Some comments follow for the reader's guidance. The book does not deal with topics such as time management, time and motion study, time series analysis, forecasting, etc., as these topics, popular though they are, do not deal with the time dimension itself as a problematic or researchable phenomenon. The writers on time management are concerned with the most efficient use and allocation of time on the part of individuals. This is the notion of time in terms of the clock. A clock-time conception is also the basis of business and other forecasting, industrial dynamics, and much of futures research. However, to afford an idea of the evolution of time reckoning, in so far as it relates to an aspect of the phenomenon of time, a chapter on clocks and calendars has been included. Futures studies, it seems to me, needs to treat the time dimension as a problematic phenomenon, and a chapter concerning it has been included here as an explicit encouragement to interested scholars.

Furthermore, almost all the sources included here are from scholarly journals and books. Conference proceedings, dissertations, theses, working papers, and occasional papers are excluded, for the obvious reason that it makes more sense to accommodate additional research publications in their place. In any event, significant findings in proceedings and dissertations tend to find their way to the research journals and books.

All entries are categorized as belonging to one of a number of disciplines. There is an initial chapter listing the general works on the time dimension, and a concluding chapter of miscellaneous entries which did not seem to belong to any of the 19 specific disciplinary chapters. It appeared appropriate to combine anthropology and culture, philosophy and physics, psychiatry and psychoanalysis, and speech and communication. The literature on psychology is clearly vast and growing. It became necessary to eliminate just about all of the research with animals, on acoustics and auditory research, and on reaction time. Readers will hopefully understand the constraints of space and the need to make choices which can arguably be considered arbitrary. Also, one suspects that some of the entries could well be categorized differently, but it is hoped that there are not too many of such cases. All entries are of course arranged alphabetically by author or editor within each chapter, with no distinction made as between journal articles, books, and book chapters.

Researchers, students, and interested lay individuals who would like to explore the time dimension (including futures studies) may wish to contact one or more of the following organizations for professional involvement and information: (1) International Society for the Study of Time, C/o Department of English, University of Victoria, P.O. Box 1700, Victoria, B.C., Canada V8W 2Y2, (2) Association for the Social Study of Time, C/o Institute of Community Studies, 18 Victoria Park Square, Bethnal Green, London E2 9PF, UK, (3) The Project on Time and Organization, C/o College of Business and Public Administration, 225 Middlebush Hall, University of Missouri-Columbia, Columbia, Missouri 65211, U.S.A., and (4) World Future Society, 4916 St. Elmo Avenue, Bethesda, MD 20814-5089, U.S.A.

The necessarily protracted process of putting together a book of this nature has been at once fun and educational. It is my hope that this interdisciplinary guide would help in some small measure in facilitating the intermingling of different temporal mindsets among interested scholars.

Acknowledgments

I am grateful to a number of individuals who have been helpful to me in carrying out this project. As the founding Co-Director of The Project on Time and Organization, Professor Allen Bluedorn of the School of Business and Public Administration, University of Missouri at Columbia, has been very supportive of this temporal endeavor. I am particularly indebted to Professor Tom Bruneau of the Department of Communication, Radford University, who furnished me with a bibliographic listing of his own on the social and cultural aspects of time.

My particular thanks to the New York Public Library for making the special facilities of the Wertheim Study Room of The Research Libraries for a period of several months during which I was able to comprehensively search for temporal material in various disciplines.

I have been helped on occasion by some of my research assistants in the endless task of tracing elusive references, for which I am most thankful to Amit Agrawal, Steve Combs, Rick Rateike, Sarita Reddy, and Michelle Yeow. I am especially grateful to Julie Levine for her highly professional assistance in the preparation of the typescript.

1

General Works

Bergson, Henri. 1964.
Matter and Memory. Atlantic Highlands, NJ: Humanities.

Buckley, Jerome Hamilton. 1966.
The Triumph of Time. Cambridge, MA: Harvard University Press.

Buhler, Charlotte, and Fred Massarik (eds.). 1968.
The Course of Human Life: A Study of Goals in the Humanistic Perspective. New York: Springer.

Campbell, Joseph (ed.). 1983.
Man and Time. Princeton, NJ: Princeton University Press.

Carlstein, Tommy, Don Parkes, and Nigel Thrift. 1978.
Timing Space and Spacing Time, 3 Vols. Vol. 1: Making Sense of Time. Vol. 2: Human Activity and Time Geography. Vol. 3: Time and Regional Dynamics. London, UK: Arnold.

Christensen, Ronald. 1984.
Order and Time: A General Theory of Prediction. Lincoln, MA: Entropy Ltd.

Daniel, Stephen, H. 1981.
"Seventeenth-century scholastic treatments of time." Journal of History of Ideas, 42:587-606.

De Grazia, Sebastian (ed.). 1964.
Of Time, Work, and Leisure. New York: Doubleday.

Doob, Leonard W. 1971.
Patterning of Time. New Haven, CT: Yale University Press.

Eisley, Loren. 1980.
The Firmament of Time. New York: Atheneum.

Elias, Norbert. 1989.
On Time. New York: Basil Blackwell.

Elton, L.R., and H. Messel. 1978.
Time and Man. Elmsford, NY: Pergamon Press.

Evans, Daryl. 1975.
Explorations in Time. Monticello, IL: Council of Planning Librarians.

Fischer, Roland (ed.). 1967.
"Interdisciplinary perspectives of time." Annals of the New York Academy of Sciences, 138:367-915.

Flood, Raymond, and Michael Lockwood (eds.). 1987.
The Nature of Time. New York: Basil Blackwell.

Fraser, J.T. 1967.
"The interdisciplinary study of time." Annals of the New York Academy of Sciences, 138:822-847.

Fraser, J.T. 1972.
"The study of time." In J.T. Fraser, F.C. Haber, and G.H. Muller (eds.), The Study of Time, pp. 479-502. New York: Springer-Verlag.

Fraser, J.T. 1975.
Of Time, Passion and Knowledge: Reflections on the Strategy of Existence. New York: Braziller.

Fraser, J.T. 1978.
Time as Conflict: A Scientific and Humanistic Study. Basel and Stuttgart: Birkhauser Verlag.

Fraser, J.T. (ed.). 1981.
The Voices of Time: A Cooperative Survey of Man's Views of Time as Expressed by the Sciences and by the Humanities, 2nd edn. New York: George Braziller.

Fraser, J.T. 1981.
"The study of time." In J.T Fraser (ed.), The Voices of Time, pp. 582-592. New York: George Braziller.

Fraser, J.T. 1982.
The Genesis and Evolution of Time: A Critique of Interpretation in Physics. Amherst, MA: University of Massachusetts Press.

Fraser, J.T. 1987.
Time, the Familiar Stranger. Redmond, WA: Tempus Books.

Fraser, J.T. (ed.). 1989.
Time and Mind: The Study of Time VI. Madison, CT: International Universities Press.

Fraser, J.T., F.C. Haber, and G.H. Muller (eds.). 1972.
The Study of Time. New York: Springer-Verlag.

Fraser, J.T., and N. Lawrence (eds.). 1975.
The Study of Time II. New York: Springer-Verlag.

Fraser, J.T., N. Lawrence, and F.C. Haber (eds.). 1986.
Time, Science, and Society in China and the West: The Study of Time V. Amherst, MA: University of Massachusetts Press.

Fraser, J.T., N. Lawrence, and D. Park (eds.). 1978.
The Study of Time III. New York: Springer-Verlag.

Fraser, J.T., N. Lawrence, and D. Park (eds.). 1981.
The Study of Time IV. New York: Springer-Verlag.

Gardet, Louis, A.J. Gurevich, Alexis Kagame, Claude Larre, G.E.R. Lloyd, Andre Neher, Raimundo Panikkar, Germano Pattaro, and Paul Ricoeur. 1976.
Cultures and Time. Paris, France: Unesco Press.

Giedon, Sigfried. 1967.
Space, Time and Architecture. Cambridge, MA: Harvard University Press.

Gioscia, Victor. 1966.
Time Forms: Beyond Yesterday and Tommorrow. New York: Gordon and Breach.

Gold, Thomas (ed.). 1967.
The Nature of Time. Ithaca, NY: Cornell University Press.

Goldman. Steven Louis. 1982.
"Modern science and Western culture: the issue of time. History of European Ideas, 3:371-402.

Grant, J., and C. Wilson (eds.). 1980.
The Book of Time. North Pomfret, VT: David and Charles.

Greenaway, Frank (ed.). 1979.
Time and the Sciences. Paris, France: Unesco Press.

Guitton, Jean. 1966.
Man in Time. Notre Dame, IN: Notre Dame University Press.

Hinton, A.G. n.d.
The Fifth Dimension: A Scientific Theory with Biological Implications. Cambridge, UK: W. Heffer & Sons, Ltd.

Hoyle, Trevor. 1977.
Through the Eye of Time. London, UK: Panther.

Hughes, G.E. 1951.
On Having the Past All Over Again. New Zealand: Victoria University College.

Kern, Stephen. 1983.
The Culture of Time and Space 1880-1918. Cambridge, MA: Harvard University Press.

Landsberg, P.T. (ed.). 1982.
The Enigma of Time. Bristol, UK: Adam Hilger.

Lewis, Wyndham. 1927.
Time and Western Man. Boston, MA: Beacon Hill.

Machamer, Peter K., and Robert G. Turnbull (eds.). 1976.
Motion and Time, Space and Matter: Interrelations in the History of Philosophy and Science. Columbus, OH: Ohio State University Press.

MacIver, Robert Morrison. 1962.
The Challenge of the Passing Years: My Encounter With Time. New York: Simon and Schuster.

Mallman, Carlos A., and Oscar Nudler (eds.). 1986.
Time, Culture and Development. Bariloche, Argentina: Fundacion Bariloche.

Mellor, D.H. 1981.
Real Time. Cambridge, UK: Cambridge University Press.

Millikan, Robert A. 1932.
Time, Matter and Values. Chapel Hill, NC: University of North Carolina Press.

Moore, Wilbert E. 1963.
Man, Time, and Society. New York: Wiley.

Morris, Richard. 1985.
Time's Arrows: Scientific Attitudes Toward Time. New York: Simon and Schuster.

Munz, Peter. 1977.
The Shapes of Time. Middletown, PA: Wesleyan University Press.

Needham, Joseph. 1986.
Time, the Refreshing River. Atlantic Highlands, NJ: Humanities Press International.

Parsons, E. 1964.
Time Devoured: A Materialistic Discussion of Duration. London: UK: Allen and Unwin.

Patrides, C.A. 1976.
Aspects of Time. Manchester, UK: Manchester University Press.

Poulet, G. 1959.
Studies in Human Time. New York: Harper Torchbooks.

Priestly, J.B. 1964.
Man and Time. Garden City, NJ: Doubleday.

Prigogine, Ilya, and I. Stengers. 1984.
Order Out of Chaos: Man's New Dialogue with Nature. New York: Bantam Books.

Salmon, P. 1985.
Living in Time. London, UK: Dent.

Santillana, Giorgio de, and Hertha von Dechend. 1969.
Hamlet's Mill: An Essay on Myth and the Frame of Time. Boston, MA: Gambit.

Shallis, Michael. 1983.
On Time: An Investigation into Scientific Knowledge and Human Experience. New York: Schocken Books.

Sirianni, Carmen J. 1987.
Work, Time and Equality. Oxford, UK: Polity Press.

Toulmin, Stephen, and June Goodfield. 1982.
The Discovery of Time. Chicago, IL: University of Chicago Press.

Voise, W., J. Zeman, J.B. Molchanov, and I.A. Akchurin. 1978.
"The study of time in Poland, Csechoslovakia, and the Soviet Union." In J.T. Fraser, N. Lawrence, and D. Park (eds.), The Study of Time III, pp. 471-499. New York: Springer-Verlag.

Whitrow, Gerald J. 1988.
Time in History: The Evolution of Our General Awareness of Time and Temporal Perspective. New York: Oxford University Press.

Wood, Douglas Kellog. 1982.
Men Against Time: Nicolas Berdyaev, T.S. Eliot, Aldous Huxley, and C.G. Jung. Lawrence, KS: University of Kansas Press.

Yaker, Henri, Humphry Osmond, and Frances Cheek (eds.). 1971.
The Future of Time: Man's Temporal Environment. Garden City, NY: Doubleday.

Zelkind, Irving, and Joseph Sprug. 1974.
Time Research: 1172 Studies. Metuchen, NJ: Scarecrow Press.

Zeman, Jiri (ed.). 1971.
Time in Science and Philosophy: An International Study of Some Current Problems. New York: Elsevier.

2

Anthropology and Culture

Abou-Zeid, Ahmed M. 1979.
"The concept of time in peasant society." In Frank Greenaway (ed.), Time and the Sciences, pp. 119-128. Paris, France: Unesco Press.

Appadurai, Arjun. 1981.
"The past as a scarce resource." Man, 16 (N.S.):201-219.

Ayoade, John A.A. 1979.
"Time in Yoruba thought." In A. Richard (ed.), African Philosophy: An Introduction, pp. 71-90. Washington, DC: University Press of America.

Baker, E.C. 1952.
"Tribal calendars." Tanganyika Notes and Records, 33:30-33.

Bascom, William R., and J. Herskovits (eds.) 1959.
Continuity and Change in African Cultures. Chicago: University of Chicago Press.

Becker, Judith. 1979.
"Time and tune in Java." In A.L. Becker and A. Yengoyan (eds.), The Imagination of Reality, pp. 197-210. Norwood, NJ: Ablex.

Beidelman, T.O. 1963.
"Kaguru time reckoning: an aspect of the cosmology of an East African people." Southwestern Journal of Anthropology, 19:9-20.

Berkhofer, R.F. 1963.
"Space, time and culture and the new frontier." Agricultural History, 38:21-30.

Bloch, Maurice. 1977.
"The past and the present in the present." Man, 12 (N.S.):278-292.

Boas, George. 1950.
The Acceptance of Time. Berkeley, CA: University of California Press.

Bogoras, Waldemar. 1925.
"Ideas of space and time in the conception of primitive religion." American Anthropologist, 27:205-266.

Bohannan, Paul. 1953.
"Concepts of time amongst the Tiv of nigeria." Southwestern Journal of Anthropology, 9:251-262.

Bourdieu, Pierre. 1963.
"The attitude of the Algerian peasant toward time." In J. Pitt-Rivers (ed.), Mediterranan Countrymen: Essays in the Social Anthropology of the Mediterranean, pp. 55-72. Paris, France: Mouton.

Bourdieu, Pierre. 1977.
Outline of a Theory of Practice. Cambridge, UK: Cambridge University Press.

Bowden, E.A.F. 1971.
"Maori adolescent temporal values." New Zealand Journal of Educational Studies, 6:62-77.

Brandon, S.G.F. 1963.
Creation Legends of the Ancient Near East. London, UK: Hodder and Stoughton.

Bray, D.H. 1971.
"Maori Adolescent temporal values," New Zealand Journal of Educational Studies, 6:62-77.

Burman, Rickie. 1981.
"Time and socioeconomic change on Simbo, Soloman Islands." Man, 16 (N.S.):251-267.

Cipolla, Carlo M. 1977.
Clocks and Culture, 1300-1700. New York: Norton.

Cooper, John M. 1942.
"Areal and temporal aspects of aboriginal South American culture." Primative Man, 15, No. 1.

Cope, L. 1919.
"Calendars of the Indian north of Mexico." University of California Publications in American Archeology and Ethnology, 16:119-76.

Corbin, Henry. 1957.
"Cyclical time in Mazdaism and Ismailism." In Man and Time, Vol. 3, pp. 115-172. New York: Bollingen Foundation.

Deregowski, Jan B. 1972.
"Valuing time." New Society, 21(515):282-284.

Doob, Leonard W. 1978.
"Time: cultural and social anthropological aspects." In Tommy Carlstein, Don Parkes, and Nigel Thrift (eds.), Making Sense of Time, pp. 56-65. New York: Wiley.

Eickelman, Dale F. 1977.
"Time in a complex society: a Moroccan example." Ethnology, 16(1):39-55.

Eliade, Mircea. 1957.
"Sacred time and myths." In Mircea Eliade, The Sacred and the Profane. New York: Harcourt, Brace and World.

Eliade, Mircea. 1959.
Cosmos and History: The Myth of the Eternal Return. New York: Harper and Row.

Evans-Pritchard, E.E. 1936.
"Nuer time-reckoning." Africa, 12:189-216.

Fabian, Johannes. 1983.
Time and the Other: How Anthropology Makes it Object. New York: Columbia University Press.

Fortes, Meyer. 1949.
"Time and social structures: an Ashanti case study." In Meyer Fortes (ed.), <u>Social Structure: Studies Presented to A.R. Radcliffe-Brown</u>, pp. 54-84. Oxford, UK: Oxford University Press.

Friedman, Jonathan. 1985.
"Our time, their time, world time: the transformation of temporal modes." <u>Ethnos</u>, 50(3-4):168-183.

Geertz, Clifford. 1973.
"Person, time, and conduct in Bali." In Clifford Geertz, <u>The Interpretation of Cultures</u>, pp. 361-411. New York: Basic Books.

Gellner, Ernest. 1958.
"Time and theory in social anthropology." <u>Mind</u>, 67:182-202.

Givens, Douglas R. 1977.
<u>An analysis of Navajo Temporality</u>. Washington, DC: University Press of America.

Graulich, Michel. 1981.
"The metaphor of the day in ancient Mexican myth and ritual." <u>Current Anthropology,</u> (Feb):45-60.

Green, Helen Bagenstose. 1972.
"Temporal attitudes in four Negro subcultures." In J.T. Fraser, F.C. Haber, and G.H. Muller (eds.), <u>The Study of Time</u>, pp. 402-417. New York: Springer-Verlag.

Gross, Daniel R. 1984.
"Time allocation: a tool for the study of cultural behavior." <u>Annual Review of Anthropology</u>, 13:519-558.

Gross, David. 1981-82.
"Space, time and modern culture." <u>Telos</u>, 50 (Winter):415-439.

Gurevich, A.J. 1976.
"Time as a problem of cultural history." In Louis Gardet et al., <u>Cultures and Time</u>, pp. 229-245. Paris, France: Unesco Press.

Hall, Jr., Edward T. 1960.
"A microcultureal analysis of time." In Anthony F.C. Wallace (ed.), Men and Cultures, pp. 118-122. Philadelphia, PA: University of Pennsylvania Press.

Hall, Jr., Edward T. 1966.
The Hidden Dimenson. Garden City, NY: Doubleday.

Hall, Jr., Edward T. 1973.
The Silent Language. Garden City, NY: Doubleday.

Hall, Jr., Edward T. 1976.
Beyond Culture. Garden City, NY: Doubleday.

Hall, Jr., Edward T. 1983.
The Dance of Life: The Other Dimension of Time. Garden City, NY: Doubleday.

Hallowell, Irving. 1937.
"Temporal orientation in western civilization and in a pre-literate society." American Anthropologist, 39:647-670.

Harrison, S. 1982.
"Yams and the symbolic representation of time in a Spike river village." Oceania, 53 (Dec.):141-162.

Hazan, Haim. 1980.
The Limbo People: A Study of the Constitution of the Time Universe Among the Aged. London, UK: Routledge and Kegan Paul.

Hazan, Haim. 1984.
"Continuity and transformation among the aged: a study in the anthropology of time." Current Anthropology, 25(5):567-578.

Hough, W. 1893.
"Time-keeping by light and fire." American Anthropologist, 6:207.

Howe, Leopold E.A. 1981.
"The social determination of knowledge: Maurice Bloch aand Balinese Time." Man, 16 (N.S.):220-234.

Hughs, C.C. 1961.
"The concept and use of time in the middle years: the St. Lawrence Island Eskimos." In R.W. Kleermeir (ed.), Aging and Leisure. New York: Oxford University Press.

Isichei, P.A.C. 1975.
"Two perspectivess to the past: history and myth." Second Order, 4(2):11-17.

Johnson, A. 1975.
"Time allocation in a Machiguenga community." Ethnology, 14:301-310.

Jones, James M. 1988.
"Cultural differences in temporal perspectives: instrumental and expressive behaviors in time." In Joseph E. McGrath (ed.), The Social Psychology of Time: New Perspectives, pp. 21-38. Newbury Park, CA: Sage.

Kagame, Alexis. 1976.
"The empiricial apperception of time and the conception of history in Bantu thought." In Louis Gardet et al., Cultures and Time, pp. 89-116. Paris, France: Unesco Press.

Kray, Mubeccel B. 1979.
"The concept of time in rural societies." In Frank Greenaway (ed.), Time and the Sciences, pp. 129-142. Paris, France: Unesco Press.

Kubik, Gerhard. 1972.
"Oral notation of some west and central African time-line patterns." Review of Ethnology, 3(22):169-176.

Larre, Claude. 1976.
"The empirical apperception of time and the conception of history in Chinese thought." In Louis Gardet et al., Cultures and Time, pp.35-62. Paris, France: Unesco Press.

Lauer, Robert H. 1981.
Temporal Man: The Meaning and Uses of Social Time. New York: Praeger.

Leach, Edmund R. 1950.
"Primative calendars." Oceania, 20:245-62.

Leach, Edmund R. 1954.
"Primitave time-reckoning." In Charles Singer, E.J. Homyard, and A.R. Hall (eds.), A History of Technology, Vol. 1, pp. 110-127. Oxford, UK: Clarendon.

Leach, Edmund R. 1961.
"Two essays concerning the symbolic representation of time." In Edmund Leach (ed.), Rethinking Anthropology, pp. 124-143. London, UK: Athlone.

Leon-Portilla, Miguel. 1988.
Time and Reality in the Thought of the Maya, 2nd edn. Norman, OK: University of Oklahoma Press.

Levine, Robert V. 1988.
"The pace of life across cultures." In Joseph E. McGrath (ed.), The Social Psychology of Time: New Perspectives, pp. 39-60. Newbury Park, CA: Sage.

Levine, Robert V., and K. Bartlett. 1984.
"Pace of life, punctuality, and coronary heart disease in six countries." Journal of Cross-Cultural Psychology, 15:233-255.

Levine, Robert V., L. West, and H.T. Reis. 1980.
"Perceptions of time and punctuality in the United States and Brazil." Journal of Personality and Social Psychology, 38:541-550.

Maltz, D.N. 1968.
"Primitive time-reckoning as a symbolic system." Cornell Journal of Social Relations, 3:85-111.

Marschack, Alexander. 1972.
The Roots of Civilization. New York: McGraw-Hill.

Mass, Jeannette P. 1985.
"Beliefs relating to temporal dimensions held by three ethnic groups in the Fiji Islands." Psychological Studies, 30(1):53-65.

Mauss, Marcel. 1979.
Seasonal Variations of the Eskimo. London, UK: Routledge and Kegan Paul.

Maxwell, Robert J. 1972.
"Anthropological perspectives." In Henri Yaker, Humphry Osmond, and Frances Cheek (eds.), The Future of Time: Man's Temporal Environment, pp. 36-72. Garden City, NY: Doubleday.

Mead, Margaret. 1965.
"The future as the basis for establishing a shared culture." Daedalus, 94(1):135-55.

Minge-Klevana, W. 1980.
"Does labour and time decrease with industrialization? A survey of time-allocation studies." Current Anthropology, 21:279-298.

Munroe, R.H. et al. 1983.
"Time allocation in four societies." Ethnology, 22:355-70.

Murray, Harry. 1984.
"Time in the streets." Human Organization, 43(2):154-161.

Needham, Joseph. 1965.
Time and Eastern Man. Occasional Paper No. 21. Glasgow, UK: Royal Anthropological Institute.

Neher, Andre. 1976.
"The view of time and history in Jewish culture." In Louis Gardet et al., Cultures and Time, pp. 149-167. Paris, France: Unesco Press.

Nilsson, Martin P. 1920.
Primitive Time-Reckoning: A Study in the Origin and First Development of the Art of Counting Time Among Primitive and Early Culture Peoples. Oxford, UK: Oxford University Press.

Ohmuki-Tierny, E. 1969.
"Concepts of time among the Ainu of the northwest coast of Sakhalin." American Anthropologist, 71:488-92.

Panoff, Michel. 1969.
"The notion of time among the Maenge people of New Britian." Ethnology, 8:153-166.

Pepelasis, A.A. 1958-59.
"The image of the past and economic backwardness." Human Organization, 17 (Winter):19-27.

Pocock, David F. 1967.
"The anthropology of time-reckoning." In J. Middleton, (ed.), Myth and Cosmos: Readings in Mythology and Symbolism, pp. 303-314. Garden City, NY: Natural History Press.

Ricoeur, Paul. 1976.
"Introduction." In Louis Gardet et al., Cultures and Time, pp. 13-33. Paris, France: Unesco Press.

Rigby, P. 1983.
"Time, historical consciousness and the Masai." Comparitive Studies in Sociology and History, 25:428-56.

Roche, Maurice. 1988.
"Time and the critique of anthropology." Philosophy of the Social Sciences, 18(1):119-124. New York: Columbia University Press.

Sapir, Edward. 1949.
"Time perspective in aboriginal American culture." In D.G. Mandelbaum (ed.), Selected Writings of Edward Sapir, pp. 389-462. Berkeley, CA: Univesity of California Press.

Schegloff, E.A. 1968.
"Sequencing in conversational openings." American Anthropologist, 70:1075-1095.

Schelling, T.C. 1973.
"Foreward, Symposium: Time in Economic Life." Quarterly Journal of Economics, 4:624-629.

Schipper, Kristofer, and Wang Hsiu-Huei. 1986.
"Progressive and regressive time cycles in Taoist ritual." In J. T. Fraser, N. Lawrence, and F.C. Haber (eds.), Time Science, amd Society in China and the West - The Study of Time V. pp. 185-205. Amherst, MA: University of Massachusetts Press.

Schwartz, Theodore. 1975.
"Relations among generations in time-limited cultures." Ethos, 3(2):309-322.

Smith, M.F. 1982.
"Bloody time and bloody scarcity: capitalism, authority and the transformation of temporal experience in a Papua New Guinea village." American Ethnology, 9:503-518.

Smith, Marian W. 1952.
Different cultural concepts of past, present and future." Psychiatry, 15:395-400.

Smith, R.J. 1961.
"Cultural differences in the life cycle and the concepts of time." In R.W. Kleemeier (ed.), Aging and Leisure. New York: Oxford University Press.

Strate, Lance. 1986.
"Time-binding in oral cultures." Etc., 43(3):234-246.

Straus, Erwin (Walter Maximilian). 1982.
Man, Time, and World: Two Contributions to Anthropological Psychology. Pittsburgh, PA: Duquesne University Press.

Tedlock, Barbara. 1981.
Time and the Highland Maya. Albuquerque, NM: University of New Mexico Press.

Thomas, K. 1964.
"Work and leisure in pre-industrial societies." Past and Present, 29:50-66.

Thompson, Richard W., and Michael C. Robbins. 1973.
"Seasonal variations in conception in rural Uganda and Mexico." American Anthropologist, 75:676-686.

Thornton, Robert J. 1980.
Space, Time and Culture among the Iraqus of Tanzania. New York: Academic.

Wax, Murray. 1962.
"The notions of nature, man, and time of a hunting people." Southern Folklore Quarterly, 26:175-186.

Wayman, Alex. 1969.
"No time, great time, and profane time in Buddhism." In J.M. Kitagawa and C.H. Long (eds.), Myths and Symbols. pp. 47-62. Chicago, IL: University of Chicago Press.

Werner, Carol M., Lois M. Haggard, Irwin Altman, and Diana Oxley. 1988.
"Temporal qualities of rituals and celebrations: a comparison of Christmas Street and Zuni Shalako." In Joseph E. McGrath (ed.), The Social Psychology of Time: New Perspectives, pp. 203-232. Newbury Park, CA: Sage.

Werner, N. 1948.
"Primitive notions on time." In N. Werner, Comparative Psychology of Mental Development, pp. 182-190. New York: International Universities Press.

Wylie, Jonathan. 1982.
"The sense of time, the social construction of reality, and the foundations of nationhood in Dominica and the Faroe Islands." Comparative Studies in Society and Hisory, 24(3):438-466.

3

Biology

Aschoff, Jurgen (ed.). 1965.
Circadian Clocks. Amsterdam, Netherlands: North-Holland.

Aschoff, Jurgen. 1965.
"Circadian rhythms in man." Science, 148:1427-1432.

Aschoff, Jurgen (ed.). 1981.
Biological Rhythms. New York: Plenum.

Aschoff, Jurgen. 1981
"Circadian system properties." In F. Obal and G. Benedek (eds.), Advances in Physiological Sciences, Vol 18, pp. 1-17. Budapest, Hungary: Academiai Kiado.

Aschoff, Jurgen. 1984.
"Circadian timing." In John Gibbon and Lorraine Allan (eds.), Timing and Time Perception, pp. 442-468. New York: New York Academy of Sciences.

Aschoff, Jurgen, Serge Daan, and Gerard A. Groos (eds.). 1982.
Vertebrate Circadian Systems: Structure and Physiology. New York: Springer-Verlag.

Aschoff, Jurgen, H. Giedke, E. Poppel, and R. Wever. 1972.
"The influence of sleep-interruption and of sleep-deprivation on circadian rhythms in human performance."

In W.B. Colquhoun (ed.), Aspects of Human Efficiency, pp. 135-150. London, UK: English Universities Press.

Atkin, R.H. 1978.
"Time as a pattern on a multi-dimensional structure." Journal of Social and Biological Structures, 1:281-295.

Binkley, S., J.B. Riebman, and K.B. Reilly. 1978.
"Timekeeping by the pineal gland." Science, 197:1181-1183.

Blum, Harold F. 1962.
Time's Arrow and Evolution. Princeton, NJ: Princeton University Press.

Boulos, Z., A.M. Rosenwasser, and M. Terman. 1980.
"Feeding schedules and the circadian organization of behavior in the rat." Behavior and Brain Research, 1:39-65.

Boulos, Z., and M. Terman. 1980.
"Food availability and daily biological rhythms." Neuroscience and Biobehavior Review, 4:119-131.

Brown, F.A. 1965.
"A unified theory for biological rhythms." In J. Aschoff (ed.), Circadian Clocks, pp. 231-261. Amsterdam, Netherlands: North-Holland.

Brown. F.A. , J.W. Hastings, and J.D. Palmer. 1970.
The Biological Clock: Two Views. New York: Academic.

Brown, F.M., and R.C. Graeber. 1982.
Rhythmic Aspects of Behavior. Hillsdale, NJ: Erlbaum.

Brumlik, J., W. Richeson, and J. Arbit. 1967.
"The origin of certain electrical cerebral rhythms." Brain Research, 3:227-247.

Bunning, Erwin. 1967.
"Known and unknown principles of biological chronometry." Annals of the New York Academy of Sciences, 138:515-524.

Church, Russell M. 1978.
"The internal clock." In S.H. Hulse, H. Fowler and W.K. Honig (eds.), Cognitive Processes in Animal Behavior, pp. 277-310. Hillsdale, NJ: Erlbaum.

Church, Russell M. 1984.
"Properties of the internal clock." In John Gibbon and Lorraine Allan (eds.), Timing and Time Perception, pp. 566-582. New York: New York Academy of Sciences.

Cloudsley-Thompson, J.L. 1961.
Rhythmic Activity in Animal Physiology and Behavior. New York: Academic.

Cloudsley-Thompson, J.L. 1978.
"Biological clocks and the synchronizers." In J.T. Fraser, N. Lawrence, & D. Park (eds.), The Study of Time III. pp. 188-215. New York: Springer-Verlag.

Colquhoun, W.P. (ed.). 1971.
Biological Rhythm and Human Performance. New York: Academic.

Comfort, A. 1964.
Aging: The Biology of Senescence. New York: Holt, Rinehart and Winston.

Condon, William S. 1982.
"Cultural Microrhythms." In M. Davis (ed.), Interaction Rhythms: Periodicity in Communicative Behavior, pp. 53-77. New York: Human Sciences Press.

Daan, Serge, and Jurgen Aschoff. 1981.
"Short-term rhythms in activity." In Jurgen Aschoff (ed.), Biological Rhythms, pp. 491-498. New York: Plenum.

Daan, Serge, and Jurgen Aschoff. 1982.
"Circadian contributions to survival." In Jurgen Aschoff, Serge Daan, and Gerard A. Groos (eds.), Vertebrate Circadian Systems: Structure and Physiology, pp. 305-321. New York: Springer-Verlag.

Davis, F.C. 1981.
"Ontogeny of circadian rhythms." In J. Aschoff (ed.), Handbook of Behavioral Neurobiology, Vol. 4, pp. 187-190. New York: Plenum.

Dowse, H.B., and J.D. Palmer. 1972.
"The chronomutagenic effect of deuterium oxide on the

period and entrainment of a biological rhythm." Biological Bulletin, 143:513-524.

Elliott, J.A., and B.D. Goldman. 1981. "Seasonal reproduction: photoperiodism and biological clocks." In N.T. Adler (ed), Neuroendocrinology of Reproduction, pp. 377-423. New York: Plenum.

Enright, J.T. 1965. "The search for rhythmicity in biological time series." Journal of Theoretical Biology, 8:426-468.

Fischer, Roland. 1967. "The biological fabric of time." Annals of the New York Academy of Sciences, 138:440-488.

Fischer, Roland. 1981. "Biological time." In J.T. Fraser (ed.), The Voices of Time, 2nd edn, pp. 357-382. Amherst, MA: The University of Massachusetts Press.

Fischer, Roland, Frances Griffin, and Leopold Liss. 1962. "Biological aspects of time in relation to (model) psychoses." Annals of the New York Academy of Sciences, 96:44-64.

Follett, B.K., and D.E. Follett (eds.) 1981. Biological Clocks in Seasonal Reproductive Cycles. Bristol, UK: John Wright and Sons.

Fraser, J.T. (ed). 1978. "Temporal levels: sociobiological aspects of a fundamental synthesis." Journal of Social and Biological Structures, 2:339-355.

Gedda, L.B.G. 1978. Chronogenetics: The Inheritance of Biological Time. Springfield, IL: C.C. Thomas.

Goodwin, B.C. 1963. Temporal Organization in Cells: A Dynamic Theory of Cellular Control Process. New York: Academic.

Goodwin, B.C. 1967. "Biological control process and time." Annals of the New York Academy of Sciences, 138: 748-758.

Goodwin, B.C. 1972.
"Temporal order as the origin at spatial order in embryos." In J.T. Fraser, F.C. Haber, and G.H. Muller (eds.), The Study of Time, pp. 190-199. New York: Springer-Verlag.

Groos, Gerard A., and Serge Daan. 1985.
"The use of the biological clocks in time perception." In John A. Michon and Janet L. Jackson (eds.), Time, Mind, and Behavior, pp. 65-74. New York: Springer-Verlag.

Gwinner, E. 1981.
"Circannual rhythms: their dependence on the circadian system." In B.K. Follett and D.E. Follett (eds), Biological Clocks in Seasonal Reproductive Cycles, pp. 153-170. Bristol, UK: John Wright.

Haber, Francis C. 1972.
"The Darwinian revolution in the concept of time." In J.T. Fraser, F.C. Haber, and G.H. Muller (eds.), The Study of Time, pp. 383-401. New York: Springer-Verlag.

Hamner, Karl C. 1981.
"Experimental evidence for the biological clock." In J.T. Fraser (ed.), The Voices of Time, 2nd edn., pp. 281-295. Amherst, MA: Univerity of Massachusetts Press.

Hamner, Karl C., and A. Takimoto. 1964.
"Circadian rhythms and plant photoperiodism." American Naturalist, 98:295-322.

Hamner, Karl C., J.C. Finn, G.S. Sirohi, T. Hoshizaki, and B.H. Carpenter. 1962.
"The biological clock at the South Pole." Nature, 195:476-480.

Harker, Janet E. 1964.
The Physiology of Diurnal Rhythms. Cambridge, UK: Cambridge University Press.

Hastings, J.W., et al. (ed.). 1976.
The Molecular Basis of Circadian Rhythms. Berlin, FRG: Dahlem Konferenzen.

Hendricks, Jon, and M.M. Seltzer (eds.). 1986.
Aging and Time. Beverly Hills, CA: Sage.

Hilts, P. 1980.
"The clock within." Science 80, 1:61-67.

Hoagland, Hudson, 1981.
"Some biochemical considerations of time." In J.T. Fraser (ed.), The Voices of Time, pp. 312-329. Amherst, MA: University of Massachusetts Press.

Jaffe, J. 1977.
"The biological significance of Markovian communication rhythms." In S. Rosenber (ed.), Sentence Production: Development in Research and Theory. Hillsdale, NJ: Erlbaum.

Kalmus, Hans. 1971.
"Biological time scales." In Jiri Zeman (ed.), Time in Science and Philosophy, pp. 147-151. New York: Elsevier.

Kalmus, Hans. 1978.
"The measurement of biological and social changes." In J.T. Fraser, N. Lawrence, and D. Park (eds.), The Study of Time III, pp. 237-264. New York: Springer-Verlag.

Kalmus, Hans. 1981.
"Organic evolution and time." In J.T. Fraser (ed.), The Voices of Time, 2nd edn., pp. 330-352. Amherst, MA: University of Massachusetts Press.

Kreiger, D.T. (ed.), 1979.
Endocrine Rhythms. New York: Raven Press.

Lecomte Du Nouy, P. 1937.
Biological Time. New York: Macmillan.

Luce, G.G. 1971.
Biological Rhythms in Human and Animal Physiology. New York: Dover.

Maddox, G.L. (ed.). 1987.
The Encyclopedia of Aging. New York: Springer-Verlag.

Mills, J.N. 1973.
"Transmission processes between clock and manifestations." In Biological Aspects of Circadian Rhythms, pp 27-84. New York: Plenum.

Moulyn, Adrian C. 1957.
Structure, Function and Purpose: An Inquiry into the Concepts and Methods of Biology from the Viewpoint of Time. New York: Liberal Arts Press.

Moulyn, Adrian C. 1957.
"Objective time and subjective time." In Adrian C. Moulyn, Structure, Function and Purpose: An Inquiry into the Concepts and Methods of Biology from the Viewpoint of Time, pp. 83-118. New York: Liberal Arts Press.

Neumann, D. 1981.
"Tidal and lunar rhythms." In J. Aschoff (ed.), Handbook of Behavioral Neurobiology, Vol. 4: Biological Rhythms, pp. 351-380.

Palmer, J.D. 1974.
Biological Clocks in Marine Organisms. New York: Wiley.

Palmer, J.D. 1976.
An Introduction to Biological Rhythms. New York: Academic.

Palmer, J.D. 1978.
"The living clocks of marine organisms." In J.T. Fraser, N. Lawrence, and D. Park (eds.), The Study of Time III, pp. 216-235. New York: Springer-Verlag.

Pengelley, E.T. 1971.
"Annual biological clocks." Scientific American, April:72-80.

Pengelley, E.T. (ed.) 1974.
Circannual Clocks. New York: Academic.

Pittendrigh, Collin S. 1972.
"On temporal organizations in living systems." In Henri Yaker, Humphry Osmond, and Frances Cheek (eds.), The Future of Time: Man's Temporal Environment, pp. 179-218. Gerden City, NY: Doubleday.

Pittendrigh, Collin S. 1981.
"Circadian organization and the photoperiodic phenomena." In B.K. Follett and D.E. Follet (eds.), Biological Clocks in Seasonal Reproductive Cycles, pp. 1-35. Bristol, UK: John Wright and Sons.

Pittendrigh, Collin S. 1981.
"Circadian systems: general perspective." In Jurgen Aschoff (ed.), Biological Rhythms, pp. 57-80. New York: Plenum.

Pittendrigh, Collin S. 1981.
"Circadian systems: entrainment." In Jurgen Aschoff (ed.), Biological Rhythms, pp. 95-124. New York: Plenum.

Pittendrigh, Collin S., and D.H. Minis. 1964.
"The entrainment of circadian oscillations by light and their role as photoperiodic clocks." American Naturalist, 98:261-294.

Redfern, P.H., I.C. Campbell, J.A. Davies, and K.F. Martin (eds.). 1985.
Circadian Rhythms in the Central Nervous System. London, UK: VCH.

Richter, C.P. 1968.
"Inherent twenty-four hour and lunar clocks in primates." Communications in Behavior Biology, A(5):305-331.

Richter, C.P. 1975.
"Astronomical references in biological rhythms." In J.T. Fraser and N. Lawerence (eds.), The Study of Time II, pp. 39-53. New York: Springer-Verlag.

Rinberg, A., and M.H. Smolensky (eds.). 1985.
Biological Rhythms and Medicine. New York: Springer-Verlag.

Rusak, B., and I. Zucker. 1975.
"Biological rhythms and animal behavior." Animal Review of Psychology, 26:137-171.

Sanford, A.J. 1971.
"A periodic basis for perception and action." In W.P. Colquhonn (ed.), Biological Rhythm and Human Performance, pp. 197-201.

Schaltenbrand, Georges. 1975.
"Cyclic states as biological space-time fields." In J.T. Fraser and N. Lawrence (eds.), The Study of Time II, pp. 54-68. New York: Springer-Verlag.

Scheving, L.E., et al. (eds.). 1974.
Chronobiology. Tokyo, Japan: Igaku Shoin.

Sokolove, P.G., and W.N. Bushnell. 1978.
"The chi square periodogram: its utility for analysis of circadian rhythms." *Journal of Theoretical Biology*, 72:131-160.

Sollberger, A. 1965.
Biological Rhythm Research. Amsterdam, Netherlands: Elsevier.

Sollberger, A. 1967.
"Biological measurements in time, with particular reference to synchronization mechanisms." *Annals of the New York Academy of Sciences*, 138:561-599.

Strehler, Bernard L. 1977.
Time, Cells and Aging. New York: Academic.

Suda, M., O. Hayashi, and H. Nakagawa (eds.). 1979.
Biological Rhythms and their Central Mechanisms. Amsterdam, Netherlands: Elsevier/North Holland Biomedical Press.

Thorpe, J.E. (ed.). 1978.
Rhythmic Activity of Fishes. New York: Academic.

Vester, Frederic. 1979.
"Time and biology." In Frank Greenaway (ed.), *Time and the Sciences*, pp. 53-69. Paris, France: Unesco Press.

Webb, W.B. (ed.). 1982.
Biological Rhythms, Sleep, and Performance. New York: Wiley.

Winfree, Arthur T. 1980.
The Geometry of Biological Time. New York: Springer-Verlag.

4
Calendars and Clocks

Andrews, George B. 1931.
"Making the revolutionary calendar." American Historical Review, 36:515-532.

Austin, L. 1938-39.
"The seasonal and gardening calendar of the Kiriwina." Oceania, 9:237-52.

Aveni, Anthony. 1989.
Empires of Time: Calendars, Clocks, and Cultures. New York: Basic Books.

Baillie, G.H. 1951.
Clocks and Watches: An Historical Bibliography. London: UK: N.A.G. Press

Bedini, S.A. 1963.
The Scent of Time: A Study of the Use of Fire and Incense for Time Measurement in Oriental Countries. Philadelphia, PA: American Philosophical Society.

Bennett, M.F. 1974.
Living Clocks in the Animal World. Springfield, IL: Charles C. Thomas.

Breasted, J.H. 1962.
"The beginnings of time measurement and the origins of

our calender." In H. Shapley, Time and Its Mysteries, pp. 52-79. New York: Collier.

Britten, F.J. 1982.
Britten's Watch and Clockmaker's Handbook, Dictionary and Guide, 16th edn. London, UK: Methuen.

Brunton, Eric. 1979.
The History of Clocks and Watches. New York: Rizzoli.

Burlingame, Roger. 1966.
Dictator Clock: 5000 Years of Telling Time. New York: Macmillian.

Clemence, G.M. 1981.
"Time measurement for scientific use." In J.T. Fraser (ed.), The Voices of Time, pp. 401-414. Amherst, MA: University of Massachusetts Press.

Cunynghame, Sir Henry Hardinge Samuel. 1970.
Time and Clocks: A Description of Ancient and Modern Methods of Measuring Time. Detroit, MI: Singing Tree Press.

De Solla Price, Derek J. 1974.
Gears from the Greeks. Philadelphia, PA: American Philosophical Society.

Enright, J.T. 1965.
"Accurate geophysical rhythms and frequency analysis." In J. Aschoff (ed.), Circadian Clocks, pp. 31-42. Amsterdam, Netherlands: North-Holland.

Fleming, Sanford. 1886.
"Time-reckoning for the twentieth century." Smithsonian Report (1886), pp. 345-366.

Fraser, J.T., and N. Lawrence. 1975.
"Special session on timekeepers and time." The Study of Time II, pp. 363-450. New York: Springer-Verlag.

Gauquelin, Michel. 1967.
The Cosmic Clocks: From Astrology to A Modern Science. Chicago, IL: Henry Regnery.

Hafele, J.C., and Richard E. Keating. 1972.
"Around-the-world atomic clocks: predicted relativistic time gains and observed relativistic time gains." Science, 177:166-170.

Hellyer, Brian. 1974.
Man the Timekeeper. London, UK: Priory Press.

Howse, Derek. 1980.
Greenwich Time and the Discovery of Longitude. New York: Oxford University Press.

Janich, Peter. 1985.
Protophysics of Time: Constructive Foundation and History of Time Measurement. Boston, MA: Reidel.

Jespersen, James, and Jane Fitz-Randolph. 1982.
From Sundials to Atomic Clocks: Understanding Time and Frequency. New York: Dover.

King, H.C. 1978.
Geared to the Stars: The Evolution of Planetariums, Orreries, and Astronomical Clocks. Toronto, Canada: University of Toronto Press.

Leach, Edmund R. 1950.
"Primitive calendars." Oceania, 20:245-62.

Leach, Edmund R. 1954.
"Primitive time-reckoning." In C. Singer et al. (eds.), A History of Technology, Vol. 1, pp. 110-27. Oxford: Claredon Press.

Lloyd, H. Alan. 1981.
"Timekeepers: an historical sketch." In J.T. Fraser (ed.), The Voices of Time, 2nd edn., pp. 388-400. Amherst, MA: University of Massachusetts Press.

Maurice, Klaus, and Otto Mayr (eds.). 1980.
The Clockwork Universe: German Clocks and Automata, 1550-1650. New York: Neal Watson Academic Publications.

Mayer, Maurice. 1980.
The Clockwork Universe: German Clocks and Automata, 1500-1650. New York: Neal Watson Academic Publications.

Needham, Joseph, Lu Gwei-Djen, John H. Combridge, and John S. Major. 1986.
The Hall of Heavenly Records: Korean Astronomical Instruments and Clocks 1380-1780. New York: Cambridge University Press.

Needham, Joseph, Wang Ling, and Derek J. De Solla Price. 1986.
Heavenly Clockwork: The Great Astronomical Clocks of Medieval China, 2nd edn. Cambridge, UK: Cambridge University Press.

Neugebauer, D. 1942.
"The origin of the Egyptian calendar." Journal of Near Eastern Studies, 1:396-403.

Nilsson, M.P. 1920/1960.
Primitive Time-Reckoning. Lund, Sweden: C.W.K. Gleerup.

O'Neil, William. 1975.
Time and the Calendars. Sydney, Australia: Sydney University Press.

Parise, Frank (ed.). 1982.
The Book of Calendars. New York: Facts on File.

Parker, Richard A. 1971.
"The calendar and chronology." In John R. Harris (ed.), The Legacy of Egypt, pp. 13-27. London, UK: Oxford University Press.

Rawlings, A.L. 1974.
The Science of Clocks and Watches. Wakefield, UK: Embury Press.

Silverberg, Robert. 1971.
Clocks for the Ages: How Scientists Date the Past. NY: Macmillan.

Thomson, Malcolm M. 1978.
The Beginning of the Long Dash: A History of Timekeeping in Canada. Toronto, Canada: University of Toronto Press.

Watkins, Harold. 1954.
Time Counts: The Story of the Calendar. New York: Philosophical Library.

Winlock, H.E. 1940.
"The origin of the Egyptian calendar." *Proceedings of the American Philosophical Society*, 83:447-464.

Wright, Lawrence. 1969.
Clockwork Man: The Story of Time -- Its Origins, Its Uses, Its Tyranny. New York: Horizon Press.

Zerubavel, Eviatar. 1985.
The Seven-Day Circle: The History and Meaning of the Week. New York: Free Press.

Zuidema, R.T. 1977.
"The Inca calendar." In Anthony F. Aveni (ed.), *Native American Astronomy*, pp. 219-259. Austin, TX: University of Austin Press.

5
Economics

Arvan, Lanny, and Leon N. Moses. 1985.
"A model of the firm in time and space." Journal of Economic Dynamics and Control, 9(1):77-100.

Bausor, R. 1983.
"Rational-expectations hypothesis and the epistemics of time." Cambridge Journal of Economics, 7 (March):1-10.

Becker, G.S. 1965.
"A theory of the allocation of time." Economic Journal, 75:493-517.

Bender, L.D., and L.C. Parcels. 1983.
"Structural differences and the time patterm of basic employment." Land Economics, 59 (May):220-234.

Bruzelius, Nils. 1979.
The Value of Travel Time. London, UK: Croom Helm.

Cencini, Alvaro. 1984.
Time and Microeconomic Analysis of Income. New York: Martin's.

Converse, Philip, E. 1968.
"Time budgets." In David L Sills (ed.), International Encyclodedia of the Social Sciences, 16:42-47. New York: Macmillian and Free Press.

DeSerpa, A.C. 1971.
"A theory of the economics of time." Economic Journal.

Epstein, L.G., and J.A. Haynes. 1983.
"Rate of time preference and dynamic economic analysis." Journal Political Economics, 91:611-635.

Evans, A.W. 1972.
"On the theory of the valuation and allocation of time." Scottish Journal of Political Economy, February.

Feldstein, M.S. 1964.
"The social time preference discount rate in cost benefit analysis." Economic Journal, June.

Folkerst-Landau, D.F.I. 1982.
Intertemporal Planning Exchange, and Macroeconomics. Cambridge, UK: Cambridge University Press.

Garrison, R.W. 1984.
"Time and money: the universals of macroeconomic theorizing." Journal of Macroeconomics, 6(2):197-213.

Geweke, J., et al. 1983.
"Comparing alternative tests of causality in temporal systems: analytic results and experimental evidence." Journal Economet, 21 (Feb.):161-194.

Ghez, G.R., and G.S. Becker. 1975.
The Allocation of Time and Goods Over the Life-Cycle. New York: Columbia University Press.

Goodwin, P.B. 1978.
"Travel choice and time budgets." In D. Hensher and Q. Dalvi (eds.), Determinants of Travel Choice. Farnborough, UK: Saxon House.

Gronau, Reuben. 1977.
"Leisure, home production and work -- the theory of the allocation of time revisited." Journal of Political Economy, 85:1099-1123.

Heggi, I. (ed.). 1976.
Model Choice and the Value of Travel Time. Oxford, UK: Oxford University Press.

Hicks, J.R. 1976.
"Some questions of time in economics." In Anthony M. Tang, Fred M. Westfield, and James J. Worley (eds.), *Evolution, Welfare, and Time in Economics: Essays in Honor of Nicholas Georgesen-Roegen*, pp. 135-152. Lexington, MA: D.C. Heath.

Ho, T.J. 1979.
"Time costs of child rearing in the rural Philippines." *Population and Development Review*, 5:643-662.

Holmes, Ivory H. 1983.
The Allocation of Time By Women Without Family Responsibilities. Washington, DC: University Press of America.

Hood, William C. 1948.
"Some aspects of the treatment of time in economic theory." *Canadian Journal of Economics and Political Science*, 14:453-468.

Jennings, A., and Clifford Sharp. 1976.
"The value of travel time savings and transport investment appraisal." In I. Heggie (ed.), *Model Choice and the Value of Travel Time*. Oxford, UK: Oxford University Press.

Johnson, L.C., D.I. Tepas, W.P. Colquhoun, and M.J. Colligan (eds.). 1981.
The Twenty-four Hour Workday: Proceedings of a Symposium on Variations in Work-Sleep Schedules. Cincinnati, OH: U.S. Department of Health and Human Services.

Johnson, III, Sam H. 1984.
"Temporal land resource concerns and farming systems research: Chiang Mai Valley, Northern Thailand." *Land Economics*, 60(2):202-210.

Lachmann, L.M. 1959.
"Professor Shackle and the economic significance of time." *Metroeconomica*, 11:64-73.

Loasby, Brian J. 1976.
Choice, Complexity and Ignorance: An Enquiry into Economic Theory and the Practice of Decision-Making. Cambridge, UK: Cambridge University Press.

Melbin, Murray. 1978.
"The colonization of time." In T. Carlstein et al. (eds.), Human Activity and Time Geography. London, UK: Edward Arnold.

Modigliani, Franco, and Franz E. Hahn. 1955.
"Production planning over time and the nature of expectations and planning horizon." Econometrica, 23:46-66.

Mueller, E. 1984.
"The value and allocation of time in rural Botswana." Journal Development Economics, 15 (May/August):329-360.

Musgrove, Philip. 1985.
"Why everything takes 2.71728 ... times as long as expected." American Economic Review, 75:250-252.

Nelson, J.R. 1968.
"The value of travel time." In S.B. Chase (ed.), Problems in Public Expenditure Analysis. Washington, DC: Brookings Institution.

Noyes, Richard. 1980.
"The time horizon of planned social change: Part I: Why utopian movements always promise amelioration in the future." American Journal of Economics and Sociology, 39:65-77.

Noyes, Richard. 1980.
"The time horizon of planned social change: Part II: How the advocates of Social Reform may expedite their purpose through temporal calibration." American Journal of Economics and Sociology, 39:261-272.

O'Driscoll, Jr., Gerald, and Mario J. Rizzo. 1985.
The Economics of Time and Ignorance. Oxford, UK: Basil Blackwell.

Olson, Mancur, and Martin J. Bailey. 1981.
"Positive time preference." Journey of Political Economy, 89:1-25.

Pollack, Robert A., and Michael L. Wachter. 1975.
"The relevance of the household production function and its implications for the allocation of time." Journal of Political Economy, 83:255-277.

Repetto, R.C. 1971.
Time in India's Development Programmes. Cambridge, MA: Harvard University Press.

Rizzo, Mario J. (ed.). 1979.
Time, Uncertainty, Disequilibrium. Lexington, MA: D.C. Heath.

Robinson, Joan. 1980.
"Time is economic theory." Kyklos, 33:219-229.

Rosenstein-Rodan, P.N. 1934.
"The role of time in economic theory." Economica, 1:77-97.

Samuelson, Paul Anthony. 1976.
"Speeding up of time with age in recognition of life as fleeting." In Anthony M. Tang, Fred M. Westfield, and James S. Worley (eds.), Evolution, Welfare, and Time in Economics: Essays in Honor of Nicholas Georgescu-Roegen, pp. 153-170. Lexington, MA: D.C. Heath.

Schelling, T.C. 1973.
"Foreword: Symposium: Time in economic life." Quarterly Journal of Economics, 4:624-29.

Shackle, G.L.S. 1958.
Time in Economics. Amsterdam, Netherland: North Holland.

Shackle, G.L.S. 1969.
Decision, Order and Time in Human Affairs, 2nd edn. Cambridge, UK: Cambridge University Press.

Shackle, G.L.S. 1972.
Epistemics and Economics: A Critique of Economic Doctrines. Cambridge, UK: Cambridge University Press.

Sharp, Clifford. 1981.
The Economics of Time. New York: Wiley.

Szalai, Alexander. 1966.
"Differential evaluation of time budgets for comparison purposes." In Richard L. Merritt and Stein Rokkan, Comparing Nations: The Use of Quantitative Data in Cross-National Research, pp. 239-258. New Haven, CT: Yale University Press.

Szalai, Alexander. 1966.
"Trends in comparative time-budget research." _American Behavioral Scientist_, 9(9):3-8.

Tang, Anthony M., Fred M. Westfield, and James S. Worley (eds.). 1976.
Evolution, Welfare, and Time in Economics: Essays in Honor of Nicholas Georgescu-Roegen. Lexington, MA: D.C. Heath.

Thompson, E.P. 1967.
"Time, work-discipline and industrial capitialism." _Past and Present_, 38:56-97.

Uzawa, H. 1968.
"Time preference, the consumption function and optimum asset holdings." In J.N. Wolfe (ed.), _Value, Capital and Growth_, pp. 185-219. Edinburgh, UK: Edinburgh University Press.

Walker, Kathryn E., and Margaret E. Woods. 1976.
Time Use: A Measure of Household Production of Family Goods and Services. Washington, DC: American Home Economics Association.

Wallis, Robert. 1970.
"Chronopolitics: the impact of time perspectives on the dynamics of change," _Social Forces_, 49(1):102-108.

Winston, Gordon C. 1982.
The Timing of Economic Activities: Firms, Households, and Markets in Time-Specific Analysis. Cambridge, UK: Cambridge University Press.

Yeung, Y.M. and S.H.K. Yeh. 1976.
"Time budgets: extended methodology and application." _Environment and Planning A_, 8:93-107.

6
Futures Studies

Abrams, Mark. 1971.
"Mass views of the future." Futures, 3:103-15.

Ackoff, Russell L. 1974.
Redesigning the Future: A Systems Approach to Societal Problems. New York: Wiley.

Adelson, Marvin, and Samuel Aroni. 1975.
"Differential images of the future." In Harold A. Linstone and Murray Turoff (eds.), The Delphi Method: Techniques and Applications, 433-462. Reading, MA: Addison-Wesley.

Aithal, S. Krishnamoorthy. 1981.
"Literature on the future and its bearing on futures research." World Futures, 17:91-101.

Amara, Roy. 1978.
"Probing the future." In Jib Fowles (ed.), Handbook of Futures Research, pp. 41-51. Westport, CT: Greenwood.

Amara, Roy. 1984.
"New directions for futures research: setting the stage." Futures, 16:401-404.

Anderson, Gunnar. 1973.
"Methods in futures studies: a view from the theory of

science." Technological Forecasting and Social Change, 5:305-319.

Anderson, Stanford (ed.). 1968.
Planning for Diversity and Choice: Possible Futures and Their Relations to the Man-Controlled Environment. Cambridge, MA: MIT Press.

Apostol, P. 1972.
"Marxism and the structure of the future." Futures, 4:201-210.

Aristotle. 1978.
"On interpretation." In Richard M. Gale (ed.), The Philosophy of Time: A Collection of Essays, pp. 179-182. NJ: Humanities.

Backstrand, Goran. 1981.
"Sweden's secretariat for futures studies." World Futures, 17:275-297.

Bart, Pauline B. 1974.
"Why women see the future differently from men." In Alvin Toffler (ed.), Learning for Tomorrow: The Role of the Future in Education, pp. 33-55. New York: Vintage Books.

Becker, Harold S. 1984.
"Making futures research useful: the practitioner's opportunity." Futures, 16:408-417.

Beckwith, Burnham P. 1967.
The Next Five Hundred Years. Jericho, NY: Exposition-University Book.

Bell, Daniel. 1964.
"Twelve modes of prediction: a preliminary sorting of approaches in the social sciences." Daedalus, 93 (Summer):845-80.

Bell, Daniel. 1973.
The Coming of Post-Industrial Society: A Venture in Social Forecasting. New York: Basic Books.

Bell, Wendell. 1974.
"Social science: the future as a missing variable." In Alvin Toffler (ed.), Learning for Tomorrow: The Role of

the Future in Education, pp. 75-102. New York: Vintage Books.

Bell, Wendell, and James A. Mau. 1970.
"Images of the future." In J.C. McKinney, and E.A. Tiryakian (eds.), Theoretical Sociology: Perspectives and Developments, pp. 205-234. New York: Appleton-Century Crofts.

Bell, Wendell, and James A. Mau (eds.). 1971.
The Sociology of the Future: Theory, Cases, and Annotated Bibliography. New York: Russell Sage Foundation.

Bell, Wendell, and James A. Mau. 1971.
"Images of the future." In Wendell Bell and James A. Mau (eds.), The Sociology of the Future: Theory, Cases, and Annotated Bibliography, pp. 6-44. New York: Russell Sage Foundation.

Bennis, Warren G. 1970.
"A funny thing happened on the way to the future." In Frances F. Korten, Stuart W. Cook, and John I. Lacey (eds.), Psychology and the Problems of Society, pp. 431-450. Washington, DC: American Psychological Association.

Beres, Louis Rene, and Harry R. Targ. 1977.
Constructing Alternative World Futures: Reordering the Planet. Cambridge, MA: Schenkman.

Bertaux, Pierre. 1968.
"The future of man." In W.R. Ewald (ed.), Environment and Change: The Next Fifty Years, pp. 13-20. Bloomington, IN: Indiana University Press.

Bestuzhev-Lada, Igor. 1969.
"Forecasting: an approach to the problems of the future." International Social Science Journal, 21(4):526-534.

Birkenhead, Earl of. 1930.
The World in 2030. London, UK: Hooper and Stoughton.

Bjerstedt, Ake. 1979.
"Preparedness for the future as an educational goal." Educational & Psychological Interactions, No. 68 (June), pp. 28.

Bjerstedt, Ake. 1982.
"Future consciousness and the school." Didakometry & Sociometry, Vol. 14(3-4):93p.

Blankenship, Jane, and J.K. Muir. 1987.
"On imaging the future: the secular search for 'piety'." Communication Quarterly, 35 (Winter, No. 1):1-12.

Boniecki, George. 1980.
"What are the limits to man's time and space perspectives? -- Toward a definition of a realistic planning horizon." Technological Forecasting and Social Change, 17:161-175.

Boulding, Elise. 1973.
"Religion, futurism, and models of social change." The Humanist, 33(6):35-39.

Brown, Harrison, et al. 1967.
The Next Ninety Years. Pasadena, CA: California Institute of Technology.

Buchen, Irving H. 1974.
"Humanism and futurism: enemies or allies?" In Alvin Toffler (ed.), Learning for Tomorrow: The Role of the Future in Education, pp. 132-143. New York: Vintage.

Bundy, Robert (ed.). 1976.
Images of the Future: The Twenty-First Century and Beyond. Buffalo, NY: Prometheus Books.

Busch, Lawrence. 1976.
"A tentative guide to constructing the future: self-conscious millenarianism." Sociological Practice, 1(1):27-39.

Choucri, Nazli, and Thomas W. Robinson (eds.). 1978.
Forecasting in International Relations: Theory, Methods, Problems, Prospects. San Francisco: W.H. Freeman.

Clarke, A.C. 1964.
Profiles of the Future. London, UK: Gollanz.

Cole, H.S.D., et al. (eds.). 1973.
Thinking About the Future. London, UK: Chatto and Windus.

Cornish, Edward, and Associates. 1977.
The Study of the Future: An Introduction to the Art and Science of Understanding and Shaping Tomorrow's World. Washington, DC: World Future Society.

Cornish, Edward. 1977.
"Towards a philosophy of futurism." The Futurist, 11(6):380-383.

Cornish, Edward S. 1981.
"The futurist revolution in research." National Forum, 61(1):6-8, 12.

Das, T.K. 1986.
"Future orientations of decision makers and the reality of strategic thinking." In Howard F. Didsbury, Jr. (ed.), Challenges and Opportunities: From Now to 2001, pp. 40-43. Bethesda, MD: World Future Society.

Das, T.K. 1986.
The Subjective Side of Strategy Making: Future Orientations and Perceptions of Executives. New York: Praeger.

Das, T.K. 1987.
"Strategic planning and individual temporal orientation." Strategic Management Journal, 8:203-109.

DeLeon, Peter. 1984.
"Futures studies and the policy sciences." Futures, 16(6):586-593.

Dror, Yehezkel. 1968.
"The role of futures in government." Futures, 1 (Sept.):40-5.

Encel, Solomon, Pauline K. Marstand, and William Page (eds.). 1975.
The Art of Anticipation: Values and Methods in Forecasting. New York: Pica Press.

Eurich, Nell. 1974.
"The humanities face tomorrow." In Alvin Toffler (ed.), Learning for Tomorrow: The Role of the Future in Education, pp. 144-156. New York: Vintage.

Evered, Roger D. 1977.
"Interest in the future: a search for unseable measures." Futuribles, 9(4):285-302.

Evered, Roger. 1983.
"Who's talking about the future? An analysis of U.S. presidents." Technological Forecasting and Social Change, 24(1):61-77.

Flechtheim, Ossip K. 1972.
"Teaching the future." In Henri Yaker, Humphry Osmond, and Frances Cheek (eds.), The Future of Time: Man's Temporal Environment, pp. 498-507. Garden City, NY: Doubleday.

Fowles, Jib. 1977.
"The problem of values in futures research." Futuribles, 9(4):303-314.

Fowles, Jib (ed.). 1978.
Handbook of Futures Research. Westport, CT: Greenwood.

Gabor, Denis. 1964.
Inventing the Future (New York: Alfred A. Knopf).

Gerardin, Lucien A. 1979.
"Structural modeling, including temporal dimensions as an aid to study complex system governabilities and to foresee unforecastable alternative futures." Technological Forecasting and Social Change, 14(4):367-385.

Gibbs, Philip. 1928.
The Day After Tomorrow: What Is Going to Happen to the World? London, UK:: Hutchinson.

Gordon, Theodore J., and John Stover. 1976.
"Using perceptions and data about the future to improve the simulation of complex systems." Technological Forecasting and Social Change, 9:191-211.

Haavelsrud, Magnus. 1983.
"Thinking about the future at school." Prospects, 13(1):9-19.

Halley, Richard B., and Harold G. Vatter. 1978.
"Technology and the future as history: a critical review of futurism." Technology and Culture, 19(1):53-82.

Harman, Willis W. 1979.
An Incomplete Ginde to the Future. New York: Norton.

Harris, Philip R. 1985.
"Future work." Personnel Journal, 64(6):53-58.

Helmer, Olaf. 1979.
"The utility of long-term forecasting." In S. Makridakis and S.C. Wheelwright (eds.), Forecasting, pp. 141-147. Amsterdam, Netherlands: North-Holland.

Hesse, Petra. 1986.
"Children's and adolescents' fears of nuclear war: is our sense of the future dissappearing?" International Journal of Mental Health, 15(1-3):93-113.

Hoare, Carol H. 1984.
"A study of practitioners' opinions regarding desirability of selected future events, time period predictions, and choice of actions in the field of adult education." Adult Education Quarterly, 35(1):11-25.

Holloway, Clark. 1978.
"Does futures research have a corporate role?" Long Range Planning, 11(5):17-24.

Hoos, Ida R. 1974.
"Criteria for 'good' futures research." Technological Forecasting and Social Change, 6:113-132.

Hoyle, Trevor. 1977.
Seeking the Mythical Future. London, UK: Panther.

Huber, Bettina J. 1974.
"Some thoughts on creating the future." Sociological Inquiry, 44(1):29-39.

Huber, Bettina J. 1978.
"Images of the future." In Jib Fowles (ed.), Handbook of Futures Research, pp. 179-224. Westpor, CT: Greenwood.

Huber, Bettina J. 1981.
"Differing national orientations toward the future: a

comparative examination of societal characteristics and public opinion." World Futures, 17:157-194.

Ingersol, David E., and Daniel Rich. 1978.
"The one-dimensional future: conceptual and political limits of scientific futurology." Alternative Futures, 1(1):5-18.

Israeli, Nathan. 1930.
"Some aspects of the social psychology of futurism." Journal of Abnormal and Social Psychology, 25:121-132.

Israeli, Nathan. 1932.
"Wishes concerning improbable future events: reactions to future." Journal of Applied Psychology, 16:584-588.

Israeli, Nathan. 1932.
"The social psychology of time, comparative rating of emotional reactions to the past, present and future." Journal of Abnormal and Social Psychology, 27:209-213.

Israeli, Nathan. 1933.
"Group predictions of future events." Journal of Social Psychology, 4:201-222.

Israeli, Nathan. 1933-34.
"Measurements of attitudes and reactions to the future." Journal of Abnormal and Social Psychology, 28:181-193.

Israeli, Nathan. 1935.
"Distress in the outlook of Lancashire and Scottish unemployed." Journal of Applied Psychology, 19:67-69.

Jones, Thomas E. 1980.
Options for the Future: A Comparative Analysis of Policy-Oriented Forecasts. New York: Praeger.

Jouvenel, Bertrand de (ed.). 1963.
Futuribles: Studies in Conjecture (I). Geneva: Droz.

Jouvenel, Bertrand de (ed.). 1965.
Futuribles: Studies in Conjecture (II). Geneva: Droz.

Jouvenel, Bertrand de (ed.). 1967.
The Art of Conjecture. New York: Basic Books.

Jungk, Robert. 1969.
"Imagination and the future." International Social Science Journal, 21(4):557-562.

Kahn, Herman, and Anthony J. Wiener. 1967.
The Year 2000: A Framework For Speculation on the Next Thirty-Three Years. New York: Macmillan.

Khakee, Abdul, and Lars Dahlgren. 1986.
"Values in futures studies and long-term planning: two Swedish case studies." Futures, 18(1):52-67.

Linstone, Harold A. 1973.
"On discounting the future." Technological Forecasting and Social Change, 4:335-338.

Linstone, Harold A. 1977.
"Confessions of a forecaster." In Harold A. Linstone and W.H. Clive Simmonds (eds.), Futures Research: New Directions, pp. 3-12. Reading, MA: Addison-Wesley.

Linstone, Harold A. 1984.
"Futures research report card -- quantity: A; quality: C; and impact: F." Futures, 16(4):396-400.

Linstone, Harold A., and W.H. Clive Simmonds (eds.). 1977.
Futures Research: New Directions. Reading, MA: Addison-Wesley.

Linstone, Harold A., and Murray Turoff (eds.). 1975.
The Delphi Method: Techniques and Applications. Reading, MA: Addison-Wesley.

Loye, David. 1978.
The Knowable Future: A Psychology of Forecasting and Prophecy. New York: Wiley.

Mankin, Don. 1978.
Toward a Post-Industrial Psychology. New York: Wiley.

Marien, Michael. 1984.
"Future studies and policy studies: complementary fields in public affairs." Policy Studies Review, 4(1):35-42.

Maruyama, Magoroh, and Arthur M. Harkins (eds.). 1978.
Cultures of the Future. The Hague, Netherlands: Mouton.

Masini, Eleonora Barbieri. 1981.
"Philosophical and ethical foundations of future studies: a discussion." World Futures, 17:1-14.

McDanield, Michael A. 1974.
"Tomorrow's curriculum today." In Alvin Toffler (ed.), Learning for Tomorrow: The Role of the Future in Education, pp. 103-131. New York: Vintage Books.

McHale, John. 1969.
The Future of the Future. New York: George Braziller.

McHale, John. 1978.
"Time and the future sense." Technological Forecasting and Social Change, 12(1):1-12.

McHale, John. 1979.
"Time and the future sense." In Frank Greenaway (ed.), Time and the Sciences, pp. 85-101. Paris, France: Unesco Press.

Meadows, Donella H., Dennis L. Meadows, Jorgen Randers, and William W. Behrens III. 1972.
The Limits to Growth: A Report for the Club of Rome's Project on the Predicament of Mankind. New York: Universe.

Molnar, Thomas. 1973.
"Elements for a critique of futurology." The Georgia Review, 27(4):560-565.

Moore, Wilbert E. 1966.
"The utility of utopias." American Sociological Review, 31:765-72.

Moore, Wilbert E. 1967.
"Forecasting the future." In W.E. Moore (ed.), Order and Change: Essays in Comparative Sociology. New York: Wiley.

Morrison, James L., William L. Renfro, and Wayne I. Boucher. 1983.
Applying Methods and Techniques of Futures Research. San Francisco, CA: Jossey-Bass.

Morrison, James L., William L. Renfro, and Wayne I. Boucher. 1984.

Futures Research and the Strategic Planning Process: Implications for Higher Education. Washington, DC: Association for the Study of Higher Education.

Muller, Herbert J. 1974.
Uses of the Future. Bloomington, IN: Indiana University Press.

Nanus, Burt. 1984.
"Futures research -- stage three." Futures, 16(4):405-407.

Ornauer, H. (ed.). 1976.
Images of the World in the Year Two Thousand: A Comparative Ten-Nation Study. The Hague, Netherlands: Mouton.

Partridge, Ernest. 1980.
"Why care about the future?" Alternative Futures, 3(4):77-91.

Piganiol, Pierre. 1969.
"Introduction: futurology and prospective study." International Social Science Journal, 21(4):515-525.

Polak, Fred L. 1961.
The Image of the Future. New York: Oceana Publications.

Poussaint, Alvin F. 1974.
"The black child's image of the future." In Alvin Toffler (ed.), Learning for Tomorrow: The Role of the Future in Education, pp. 56-71. New York: Vintage Books.

Prehoda, Robert W. 1967.
Designing the Future: The Role of Technological Forecasting. Philadelphia, PA: Chilton Books.

Raskin, Marcus. 1977.
"Futurama." Partisan Review, 44(3):387-397.

Richardson, John M. 1984.
"The state of futures research -- personal reflections." Futures, 16(4):382-395.

Sande, Oystein. 1972.
"Future consciousness." Journal of Peace Research, 9(3):271-278.

Shane, Harold G., and June Grant Shane. 1974.
"Educating the youngest for tomorrow." In Alvin Toffler (ed.), Learning for Tomorrow: The Role of the Future in Education, pp. 181-196. New York: Vintage Books.

Shippey, T.A. 1977.
"The traps of time and space." Futures, 9:234-240.

Singer, Benjamin D. 1974.
"The future-focused role-image." In Alvin Toffler (ed.), Learning for Tomorrow: The Role of the Future in Education, pp. 19-32. New York: Vintage Books.

Spencer, Martin E. 1982.
"Social science and the consciousness of the future." Theory and Society, 11:682-712.

Stanton, Roger R. 1979.
"Future organizations: a model of structural response to organizational environment." Technological Forecasting and Social Change, 15(3):217-240.

Stine, G. Harry. 1983.
The Hopeful Future. New York: Macmillan.

Strudler, Harold L. 1974.
"Educational futurism: perspective or discipline?" In Alvin Toffler (ed.), Learning for Tomorrow: The Role of the Future in Education, pp. 173-178. New York: Vintage Books.

Taylor, Keith. 1977.
"From prophecy to prediction: Saint-Simon and the conquest of the future." Futuribles, 9(1):58-64.

Thompson, Thomas H. 1978.
"Are we obliged to future others?" Alternative Futures, 1(1):29-37.

Thomson, Sir George. 1960.
The Foreseeable Future, rev. ed. Cambridge, UK: Cambridge University Press.

Toch, Hans H. 1958.
"The perception of future events: case studies in social prediction." Public Opinion Quarterly, 22 (Spring):57-66.

Toffler, Alvin. 1969.
"Value impact forecaster -- a profession of the future." In K. Baier and N. Rescher (eds.), Values and the Future, pp. 1-30. New York: Free Press.

Toffler, Alvin. 1971.
Future Shock. New York: Bantam.

Toffler, Alvin (ed.). 1974.
Learning For Tomorrow: The Role of the Future in Education, pp. 3-18. New York: Vintage Books.

Toffler, Alvin. 1981.
The Third Wave. New York: Bantam.

Valaskakis, Kimon. 1976.
"The chronospace: a proposed methodology for future studies." Futures, 8(4):320-330.

Wells, H.G. 1913.
The Discovery of the Future. New York: B.W. Huebsch.

Wilson, Kenneth D. 1978.
"Forecasting futures." Society, 15, 2 (112):22-27.

Winthrop, Henry. 1969.
"Social costs and studies of the future." Futures, 1 (Dec.):488-99.

Wren-Lewis, John. 1974.
"Educating scientists for tomorrow." In Alvin Toffler (ed.), Learning for Tomorrow: The role of the Future in Education, pp. 157-172. New York: Vintage Books.

Wrong, Dennis H. 1974.
"On thinking about the future." American Sociologist, 9(1):26-31.

7
Geography

Allan, W. 1971.
"Space-time budgets and activity studies in urban geography and planning." Environment and Planning, 3:353-368.

Bullock, N., P. Dickens, M. Shapcott, and P. Steadman. 1974.
"Time budget and models of urban activity patterns." Social Trends, 5:45-63.

Carlstein, Tommy, D. Parkes, and N. Thrift (eds.). 1978.
Human Activity and Time Geography: Timing Space and Spacing Time II. London, UK: Edward Arnold.

Carlstein, Tommy, D. Parkes, and N. Thrift (eds.). 1978.
Time and Regional Dynamics: Timing Space and Spacing Times III. London, UK: Edward Arnold.

Chapin, Jr., F.S. 1974.
Human Activity Patterns in the City: Things People Do in Time and in Space. New York: Wiley.

Gregson, Nicky. 1986.
"On duality and dualism: the case of structuration and time geography." Progress in Human Geography, 10(2):184-205.

Knopf, Adolph. 1949.
"The geological records of time." In Time and its Mysteries, Series III, pp. 33-59. New York: New York University Press.

Lawton, R. 1983.
"Space, place and tome." Geography, LXVIII (June):193-207.

Philbrick, A.K. 1982.
"Hierarchical nodality in geographical time-space." Economic Geography, 58 (January):1-19.

Szalai, Alexander (ed.). 1972.
The Use of Time: Daily Activities of Urban And Suburban Populations in Twelve Countries. The Hague, Netherlands: Mouton.

Thrift, Nigel. 1977.
"Time and theory in human geography, Part I." Progress in Human Geography, 1:65-101.

Thrift, Nigel. 1981.
"Owners' time and own time: the making of a capitalist time consciousness, 1300-1880." In Space, Time, and Geography, pp. 56-84. Lund, Sweden: GWK Gleerup.

8
Geology

Eicher, Don L. 1976.
Geological Time, 2nd edn. Englewood Cliffs, NJ: Prentice-Hall.

Fairbridge, R.W. 1967.
"Geological and cosmic cycles." *Annals of the New York Academy of Sciences*, 138:433-439.

Gould, Stephen Jay. 1987.
Time's Arrow, Time's Cycle: Myth and Metaphor in the Discovery of Geological Time. Cambridge, MA: Harvard University Press.

Kirkaldy, J.F. 1977.
Geological Time. Edinburgh, UK: Oliver and Boyd.

9
History

Alexander, Hubert G. 1945.
Time as Dimension and History. Albuquerque, NM: University of New Mexico Press.

Allan, George. 1985.
The Importances of the Past: A Meditation on the Authority of Tradition. New York: State University of New York Press.

Ariotti, P.E. 1972.
"Concept of time in late antiquity." International Philosophical Quarterly, 12:526-552.

Ariotti, P.E. 1975.
"The Concept of time in Western antiquity." In J.T. Fraser and N. Lawrence (eds.), The Study of Time II, pp. 69-80. New York: Springer-Verlag.

Brandon, S.G.F. 1965.
History, Time and Deity. Manchester, UK: Manchester University Press.

Braudel, F. 1958.
"Time, history and the social sciences." In F. Stern (ed.), The Varieties of History. New York: Meridian Books.

Britsch, Todd A. 1980.
"History and the future." Alternative Futures, 3(3):46-54.

Carlstein, Tommy. 1982.
Time Resources, Society and Ecology: On the Capacity for Human Interaction in Space and Time, Vol. 1: Preindustrial Societies. London, UK: George Allen & Unwin.

d'Azavedo, Warren L. 1962.
"Uses of the past in Gola discourse." Journal of African History, 3(1):11-34.

Florovsky, George. 1969.
"The study of the past." In Ronald H. Nash (ed.), Ideas of History, vol. 2, pp. 351-369. New York: Dutton.

Gardet, Louis. 1976.
"Moslem views of time and history: an essay in cultural typology." In Louis Gardet et al., Cultures and Time, pp. 197-227. Paris, France: Unesco Press.

Haber, Francis C. 1978.
The Age of the World, 2nd edn. Westport, CT: Greenwood.

Harrison, Edward. 1985.
The Masks of the Universe. New York: Macmillan.

Jahoda, G. 1963.
"Children's concepts of time and history." Educational Review, 15:87-104.

Koselleck, Reinhart. 1985.
Futures Past: On the Semantics of Historical Time. Cambridge, MA: MIT Press.

Kracauer, S. 1966.
"Time and History." In History and the Concept of Time, pp. 65-78. Middletown, CT: Wesleyan University Press.

Kubler, George. 1962.
The Shape of Time: Remarks on the History of Things. New Haven, CT: Yale University Press.

Kubler, George. 1967.
"Style and the representation of historical time." Annals of the New York Academy of Sciences, 138:849-855.

Lynd, S. 1968.
"Historical past and existential present." In T. Roszak (ed.), The Dissenting Academy, pp. 101-109. New York: Pantheon Books.

Marcus, John T. 1961.
"Time and the sense of history: West and East." Comparative Studies in Society and History, 3:123-139.

Momigliano, Arnoldo. 1966.
"Time in ancient historiography." In George H. Nadel (ed.), History and the Concept of Time, pp. 1-23. Middletown, CT: Wesleyan University Press.

Namenwirth, J. Zvi. 1973.
"Wheels of time and the interdependence of value change in America." Journal of Interdisciplinary History, 3:649-683.

Shotwell, James T. 1949.
"Time and historical perspective." Time and its Mysteries, Series III, pp. 63-91. New York: New York University Press.

Starr, C.G. 1966.
"Historical and philosophical time." In History and the Concept of Time, pp. 24-35. Middletown, CT: Wesleyan University Press.

Thornton, Stephen J., and Ronald Vukelich. 1988.
"Effects of children's understanding of time concepts on historical understanding." Theory and Research in Social Education, 16(1):69-82.

Voise, W. 1975.
"On historical time in the works of Leibniz." In J.T. Fraser and N. Lawrence (eds.), The Study of Time II, pp. 114-121. New York: Springer-Verlag.

10

Linguistics

Adelaar, Mascia, and Vincenzo Lo Cascio. 1986. "Temporal relation, localization and direction." In Vincenzo Lo Cascio and Co Vet (eds.), Temporal Structure in Sentence and Discourse, pp. 251-297. Dordrecht, Netherlands: Foris.

Bertinetto, Pier Marco. 1986. "Intrinsic and extrinsic temporal references: on restricting the notion of 'reference time'." In Vincenzo Lo Cascio and Co Vet (eds.), Temporal Structure in Sentence and Discourse, pp. 41-78. Dordrecht, Netherlands: Foris.

Bull, W.E. 1971. Time, Tense, and the Verb. Berkeley, CA: University of California Press.

Gabbay, Dov, and Julius Moravcsik. 1980. "Verbs, events, and the flow of time." In Christian Rohner (ed.), Time, Tense, and Quantifiers, pp. 59-83. Tubingen: Niemeyer.

Kamp, Hans. 1979. "Events, instants, and temporal reference." In R. Bauerle, U. Egli, and A. von Stechow (eds.), Semantics From Different Points of View, pp. 376-418. New York.

Kamp, Hans. 1980.
"Some remarks on the topic of change, Part I." In Christian Rohrer (ed.), _Time, Tense, and Quantifiers_, pp. 135-179. Tubingen: Niemeyer.

Kasher, Asa, and Ruth Manor. 1980.
"Simple present tense." In Christian Rohrer (ed.), _Time, Tense and Quantifiers_, pp. 315-328. Tubingen: Niemeyer.

Liberman, M., and A. Prince. 1977.
"On stress and linguistic rhythm." _Linguistic Inquiry_, 8:249-336.

Lo Cascio, Vincenzo, and Co Vet (eds.). 1986.
Temporal Structure in Sentence and Discourse. Dordrecht, Netherlands: Foris.

McCawley, J. 1971.
"Tense and time in English." In J. Ch. Fillmore and D.T. Langendoen (eds.), _Studies in Linguistic Semantics_, pp. 96-113. New York: Holt, Rinhehart and Winston.

Mourelatos, A. 1978.
"Events, processes and states." _Linguistics and Philosophy_, 2:415-435.

Partee, B. 1984.
"Temporal and nominal anaphora." _Linguistics and Philosophy_, 7:243-286.

Rigotti, Francesca. 1986.
"Metaphors of time." _Etc._, 43(2):157-168.

Riviere, C. 1980.
"Tense, aspect and time location." _Linguistics_, 18(1-2): 105-135.

Rohrer, Christian (ed.) 1980.
Time, Tense, and Quantifiers: Proceedings of the Stuttgart Conference on the Logic of Tense and Quantification. Tubingen: Niemeyer.

Smith, Carlota S. 1978.
"The syntax and interpretation of temporal expressions in England." _Linguistics and Philosophy_, 2(1):43-101.

Smith, Carlota S. 1980.
"Temporal structures in discourse." In Christian Rohrer (ed.), *Time, Tense, and Quantifiers*, pp. 355-374. Tubingen: Niemeyer.

Smith, Carlota S. 1981.
"Semantic and syntactic constraints on temporal interpretation." In Ph. Tedeschi and A. Zaenen (eds.), *Tense and Aspect*, pp. 213-237. London, UK.

Whorf, Benjamin Lee. 1950.
"An American Indian model of the universe." *International Journal of American Linguistics*, 16:67-72.

11

Literature

Alkon, P.K. 1979.
Defoe and Fictional Time. Atlanta, GA: University of Geargia Press.

Church, M. 1978.
"Fiction: the language of time -- Thomas Mann and James Joyce". In J.T. Fraser, N. Lawrence, and D. Parks (eds.), The Study of Time III, pp. 500-511. New York: Springer-Verlag.

De Romilly, Jacqueline. 1968.
Time in Greek Tragedy. Ithaca, NY: Cornell University Press.

Engstrom, Alfred Garvin. 1975.
"Thyme, the chef of seasoners: on time and its images in literature." In Alfred Garvin Engstrom, Darkness and Light, pp. 129-151. University, MS: Romance Monographs.

Fawkner, H.W. 1984.
The Timescapes of John Fowles. London, UK: Associated University Press.

Ford, George H. 1978.
"The eternal movement: D.H. Lawrence's The Rainbow and Women in Love." In J.T. Fraser, N. Lawrence, and D. Park (eds.), The Study of Time III, pp. 512-536. New York: Springer-Verlag.

Ford, George H. 1986.
"Humanities and the experiences of time." In J.T. Fraser, N. Lawrence, and F.C. Haber (eds.), Time, Science, and Society in China and the West: The Study of Time V, pp. 141-148. Amherst, MA: University of Massachusetts Press.

Gale, Steven H. 1984.
"Nature half created, half perceived: time and reality in Harold Pinter's later plays." Journal of Evolutionary Psychology, 5(3-4):196-204.

Harmon, William. 1977.
Time in Ezra Pound's Work. Chapel Hill, NC: University of North Carolina Press.

Higdon, D.L. 1977.
Time in English Fiction. New York: Macmillan.

Jackson, Robert. 1988.
The Dismantling of Time in Contemporary Poetry. Tuscaloasa, AL: University of Alabama Press.

Kastan, David Scott. 1982.
Shakespeare and the Shapes of Time. Hanover, NH: University Press of New England.

Kemp, T. Peter, and David Rasmussen (eds.). 1989.
The Narrative Path: The Later Works of Paul Ricoeur. Cambridge, MA: MIT Press.

Krey, Laura. 1946.
"Time and the novel." In William S. Knickerbocker, Twentieth Century English. New York: Philosophical Library.

Lakoff, George, and Mark Johnson. 1980.
Metaphors We Live By. Chicago, IL: University of Chicago Press.

Lynen, John F. 1969.
The Design of the Present: Essays on Time and Form in American Literature. New Haven, CT: Yale University Press.

Macey, Samuel L. 1986.
"Literary images of progress." In J.T. Fraser, N.

Lawrence, and F.C. Haber (eds.), Time, Science, and Society in China and the West: The Study of Time V, pp. 93-103. Amherst, MA: University of Massachusetts Press.

Mendilow, Adam L. 1952.
Time and the Novel. New York: Peter Nevill.

Meyerhoff, Hans. 1955.
Time in Literature. Berkeley, CA: University of California Press.

Morris, Jill. 1977.
Time and Timelessness in Virginia Woolf. Hicksville, NY: Exposition Press.

Quinones, R.J. 1972.
The Renaissance Discovery of Time. Cambridge, MA: Harvard University Press.

Quinones, R.J. 1975.
"Four phases of time and literary modernism." In J.T. Fraser and N. Lawrence (eds.), The Study of Time II, pp. 122-135. New York: Springer-Verlag.

Quinones, R.J. 1985.
Mapping Literary Modernism: Time and Development. Princeton, NJ: Princeton University Press.

Rollyson, Jr., Carl Edmund. 1984.
Uses of the Past in the Novels of William Faulkner. Ann Arbor, MI: UMI Research Press.

Sastri, S. 1963.
"Conception of time in post-vedic Sanskrit literature." In Essays in Indology, pp. 149-204. Delhi, India: Meharchand Lachhmandasa.

Soulsby, Marlene Pilarcik. 1989.
"Mind and time: a comparative reading of Haiku, Kafka, and Le Guin." In J.T. Fraser (ed.), Time and Mind: The Study of Time VI. Madison, CT: International Universities Press.

Stutterheim, C.F.P. 1981.
"Time in Language and Literature." In J.T. Fraser (ed.), The Voices of Time, 2nd edn., pp. 163-179. Amherst, MA: University of Massachusetts Press.

Survant, Joe. 1984.
"Narrative time and the spatial metaphor: phenomenology as problem and cure." Journal of Evolutionary Psychology, 5(1-2):36-42.

Sypher, Wylie. 1976.
The Ethic of Time: Structures of Experience in Shakespeare. New York: Seabury Press.

Turner, Frederick. 1971.
Shakespeare and the Nature of Time: Moral and Philosophical Themes in Some Plays and Poems of William Shakespeare. Oxford, UK: Clarendon Press.

Turner. Frederick. 1978.
"Poiesis: time and artistic discourse." In J.T. Fraser, N. Lawrence, and D. Park (eds.), The Study of Time III, pp. 614-633. New York: Springer-Verlag.

Turner, Frederick. 1986.
"Space and time in Chinese verse." In J.T. Fraser, N. Lawrence, and F.C. Haber (eds.), Time, Science, and Society in China and the West: The Study of Time V, pp. 241-251. Amherst, MA: University of Massachusetts Press.

Williams, Mary Ellen. 1979.
A Vast Landscape: Time in the Novels of Thornton Wilder. Pocatello, ID: Idaho State University Press.

12

Management and Organizational Studies

Ackoff, Russell L. 1981.
Creating the Corporate Future. New York: Wiley.

Bjorkman, Mats. 1984.
"Decision making, risk taking and psychological time: review of empirical findings and psychological theory." Scandinavian Journal of Psychology, 25(1):31-49.

Bluedorn, Allen C., and Robert B. Denhardt. 1988.
"Time and organizations." Journal of Management, 14(2):299-320.

Bottger, Preston C. 1984.
"Expertise and air time as bases of actual and perceived influence in problem-solving groups." Journal of Applied Psychology, 69(2):214-221.

Bouwen, Rene. 1977.
"Anticipation and realization: attitudes and buying plans in the future time orientation of consumer behavior." Psychologica Belgica, 17(2):113-134.

Carnevale, Peter J., and Donald E. Conlon. 1988.
"Time pressure and strategic choice in mediation." Organizational Behavior and Human Decision Processes, 42(1):111-133.

Carnevale, Peter J., and Edward J. Lawler. 1986. "Time pressure and the development of integrative agreements in bilateral negotiations." Journal of Conflict Resolution, 30(4):636-659.

Clark, Peter A. 1977. "The plurality of time reckoning systems in modern industrial organizations." In R. Stoffle (ed.), Human Aspects of Industrialization. London, UK: Van Gorcum.

Clark, Peter A. 1978. "Temporal inventories and time structuring in large organizations." In J.T. Fraser, N. Lawrence, and D. Park (eds.), The Study of Time III, pp. 391-416. New York: Springer-Verlag.

Clark, Peter A. 1985. "A review of the theories of time and structure for organizational sociology." Research in the Sociology of Organizations, 4:35-79.

Cummings, L.L. 1982. "Organizational behavior." Annual Review of Psychology, 33:541-579.

Darasse, Claire. 1988. "Time perspective and organisational involvement." Applied Psychology: An International Review, 37(4):411-425.

Das, T.K. 1986. "Future orientations of decision makers and the reality of strategic thinking." In Howard F. Didsbury, Jr. (ed.), Challenges and Opportunities: From Now to 2001, pp. 40-43. Bethesda, MD: World Future Society.

Das, T.K. 1986. The Subjective Side of Strategy Making: Future Orientations and Perceptions of Executives. New York: Praeger.

Das, T.K. 1987. "Strategic planning and individual temporal orientation." Strategic Management Journal, 8:203-209.

Davis, Stanley M. 1982.
"Transforming organizations: the key to strategy in context." Organizational Dynamics, Winter:64-80.

Davis, Stanley M. 1987.
Future Perfect. Reading, MA: Addision-Wesley.

Dean, Douglas, John Mihalasky, Sheila Ostrander, and Lyn Schroeder. 1974.
Executive ESP. Englewood Cliffs, NJ: Prentice-Hall.

Ebert, Ronald, J., and DeWayne Piehl. 1973.
"Time horizon: a concept for management." California Management Review, 15(4):35-41.

Eden, Colin. 1977.
"Modelling the influence of decision makers on the future." Futures, 9 (August):272-284.

Eden, Colin, Sue Jones, David Sims, and Hugh Gunton. 1979.
"Images into models: the subjective world of the policy maker." Futures, 11:56-62.

Edmunds, Stahrl W. 1982.
"The role of futures studies in business strategic planning." Journal of Business Strategy, 3(2):40-46.

English, J. Morley. 1978.
"A perceptual-time scale for determination of a discount fucntion." In Cees van Dam (ed.), Trends in Financial Decision-Making, pp. 229-247. Leiden, Netherlands: Martinus Nijhoff.

Epton, S.R. 1972.
"The understimation of project duration: an explanation in terms of a time-horizon." R&D Management, 2(3):141-142.

Evered, Roger D. 1977.
"Personal scenarios: an empirical study of their relation to individual performance and to organizational activism." Human Relations, 30(11):1057-1069.

Ewing, David W. 1972.
"The time dimension." In David W. Ewing (ed.), Long-Range Planning for Management, 3rd edition, pp. 439-450. New York: Harper & Row.

Fischhoff, Baruch. 1980.
"For those condemned to study the past: reflections on historical judgement." In R.A. Shweder and D.W. Fiske (eds.), New Directions for Methodology of Behavioral Science: Fallible Judgement in Behavioral Research. San Franciso, CA: Jossey-Bass.

Friedman, Yoram, and Eli Segev. 1976.
"Horizon for strategic planning." Long Range Planning, 9(5):84-89.

Frye, R., and T. Stritch. 1964.
"Effect of timed vs. non-timed discussion upon measures of influence and change in small groups." Journal of Social Psychology, 63:139-143.

Gafni, Amiram, and George W. Torrance. 1984.
"Risk attitude and time preference in health." Management Science, 30(4):440-451.

Gersick, Connie J.G. 1988.
"Time and transition in work teams: toward a new model of group development." Academy of Management Journal, 31:9-41.

Gersick, Connie J.G. 1989.
"Marking time: predictable transitions in task groups." Academy of Management Journal, 32:274-309.

Gherardi, Silvia, and Antonio Strati. 1988.
"The temporal dimension in organizational studies." Organization Studies, 9(2):149-164.

Glennan, Jr., Thomas K. 1972.
"Evaluating federal manpower programs: notes and observations." In Pter H. Rossi and Walter Williams (eds.), Evaluating Social Programs: Theory, Practice, and Politics, pp. 187-220. New York: Seminar Press.

Goodman, Richard Alan. 1973.
"Environmental knowledge and organizational time horizon: some functions and dysfunctions." Human Relations, 26:215-226.

Goodman, Richard Alan. 1978.
Temporary Systems. New York: Praeger.

Goodman, Richard Alan, and Lawrence Peter Goodman. 1976. "Some management issues in temporary systems: a study of professional development and manpower -- the theater case." Administrative Science Quaterly, 21:494-501.

Graham, Robert J. 1981. "The role of perception of time in consumer research." Journal of Consumer Research, 7:335-342.

Graham, Robert J. 1982. "Perception of time in consumer research: rejoinder." Journal of Consumer Research, 9:119-120.

Gulick, Luther. 1987. "Time and public administration." Public Administration Review, 47:115-119.

Gupta, Shalini, and L.L. Cummings. 1986. "Perceived speed of time and task affect." Perceptual and Motor Skills, 63(2,Pt.2):971-980.

Hermann, C. 1972. "Threat, time, and surprise: a simulation of international crises." In C. Hermann (ed.), International Crises: Insights from Behavioral Research, pp. 187-211. New York: Free Press.

Heroux, Lise, Michel Laroche, and K. Lee McGown. 1988. "Consumer product label information processing: an experiment involving time pressure and distraction." Journal of Economic Psychology, 9(2):195-214.

Hill, J.M.M. 1956. "The time-span of discretion in job analysis." Human Relations, 9:295-323.

Hoffman, L.R., and R. O'Day. 1979. "The process of solving reasoning and value problems." In L.R. Hoffman (ed.), The Group Problem-Solving Process: Studies of a Valence Model, pp. 66-77. New York: Praeger.

Hornik, Jacob. 1984. "Subjective vs. objective time measures: a note on the perception of time in consumer behavior." Journal of Consumer Research, 11(1):615-618.

Huber, George P. 1985.
"Temporal stability and response-order biases in participant descriptions of organizational decisions." Academy of Management Journal, 28(4):943-950.

Isenberg, D.J. 1981.
"Some effects of time-pressure on vertical structure and decision-making accuracy in small groups." Organizational Behavior and Human Performance, 27:119-134.

Jacoby, Jacob, George J. Szybillo, and Carol Kohn Berning. 1976.
"Time and consumer behavior: an interdisciplinary overview." Journal of Consumer Research, 2:320-339.

Jaques, Elliott. 1956.
Measurement of Responsibility. Cambridge, MA: Harvard University Press.

Jaques, Elliott. 1961.
Equitable Payment. London, UK: Heinemann.

Jaques, Elliott. 1964.
Time-Span Handbook. London, UK: Heinemann.

Jaques, Elliott. 1979.
"Taking time seriously in evaluating jobs." Harvard Business Review, September-October:124-132.

Johnson, L.L., D.I Tepas, W.P. Colquhoun, and M.J. Colligan (eds.). 1981.
Biological Rhythms, Sleep and Shift Work. Jamaica, NY: SP Medical and Scientific Books.

Kantrow, Alan (ed.), 1986.
"Why history matters to managers." Harvard Business Review, January-February, 64(1):81-88.

Katz, Ralph. 1980.
"Time and work: toward an intergrated perspective." Research in Organizational Behavior, 2:81-127.

Kimberly, John R., et al. 1980.
The Organizational Life Cycle: Issues in the Creation, Transformation and Decline of Organizations. San Francisco, CA: Jossey-Bass.

Kimberly, John R. 1980.
"Data aggregation in organizational research: the temporal dimension." Organization Studies, 1:367-377.

Kolaja, Jiri. 1981.
"Anticipation of the future." International Review of Sociology, 17(2-3):163-172.

Lowenstein, George F. 1988.
"Frames of mind in intertemporal choice." Management Science, 34:200-214.

Makridakis, Spyros, and Steven C. Wheelwright. 1981.
"Forecasting an organization's futures." In Paul C. Nystrom, and William H. Starbuck (eds.), Handbook of Organizational Design, Vol. 1, pp. 122-138. New York: Oxford University Press.

McGrath, Joseph E, and Nancy L. Rotchford. 1983.
"Time and behavior in organization." Research in Organizational Behavior, 5:57-101.

McGrath, Joseph E, and Nancy L. Rotchford. 1983.
"Coping with time: time, stress, and coping in organizational settings." In Barry M. Staw and Larry L. Cummings (eds.), Research in Organizational Settings.

McGrath, Joseph E. 1988.
"Toward a time-based theory of functional groups." In R.P. McGlynn (ed.), Groups and Organizations. Lubbock, TX: Texas Tech University Press.

Miller, Eric J. 1959.
"Technology, territory, and time: the internal differentiation of complex production systems." Human Relations, 12:243-272.

Miller, Ernest C. 1977.
"Time-span of discretion and human resources management: an interview with Elliott Jaques." Personnel, 54:47-58.

Miller, G.A., et al. 1960.
Plans and the Structure of Behavior. New York: Holt, Rinehart and Winston.

Mitchell, Deborah J., J. Edward Russo, and Nancy Pennington. 1989.

"Back to the future: temporal perspective in the explanation of events." Journal of Behavioral Decision Making, 2:25-38.

Moore, Danny L., Douglas Hausknecht, and Kanchana Thamodaran. 1986.
"Time compression, response opportunity, and persuasion." Journal of Consumer Research, 13(1):85-99.

Morello, Gabriele. 1976.
"Time orientation in management students: the European case." Management International Review, 16(1):59-69.

Mueller, Guenter, and James B. Smith. 1984.
"Six 'commandments' for successful futures studies for corporate planning." Journal of Business Strategy, 5(2):88-92.

Neelameghan, A. 1979.
"Expressions of time in information science and their implications." In Frank Greenaway (ed.), Time and the Sciences, pp. 103-118. Paris, France: Unesco Press.

Neustadt, Richard E., and Ernest R. May. 1986.
Thinking in Time: The Uses of History for Decision-Makers. New York: Free Press.

Olsen, Jr., E. Odgers. 1976.
"Roles of time in the neoclassical theory of the firm." Journal of Business, 49:541-554.

Perelman, Lewis J. 1980.
"Time in system dynamics." In Augusto A. Legasto, Jr., Jay W. Forrester, and James M. Lyneis (eds.), System Dynamics, pp. 75-89. Amsterdam, Netherlands: North-Holland.

Pessemier, Edgar A., and Moshe Handelsman. 1984.
"Temporal variety in consumer behavior." Journal of Marketing Research, 21(4):435-444.

Phillips, Jr., Steve, and Mohamed I. Dessouky. 1977.
"Solving the project time/cost tradeoff problem using the minimal cut concept." Management Science, 24:393-400.

Pruitt, D., and J. Drews. 1969.
"The effect of time pressure, time elapsed, and the

opponent's concession rate on behavior in negotiation." Journal of Experimental Social Psychology, 5:43-60.

Roberts, Karlene H., and Nakiye A. Boyacigiller. 1984. "Cross-national organizational research: the grasp of the blind men." Research in Organizational Behavior, 6:423-475.

Roth, Julius A. 1963. Timetables: Structuring the Passage of Time in Hospital Treatment and Other Careers. Indianapolis, IN: Bobbs-Merrill.

Rothstein, Howard G. 1986. "The effects of time pressure on judgement in multiple cue probability learning." Organization Behavior in Human Decision Processes, 37(1):83-92.

Roy, D. 1960. "'Banana time,' job satisfaction and informal interaction." Human Organization, 18(4):158-168.

Savickas, Mark L., S. Marc Silling, and Steven Schwartz. 1984. "Time perspective in vocational maturity and career decision making." Journal of Vocational Behavior, 25(3):258-269.

Savitt, Ronald. 1986. "Time, space, and competition: formulations for the development of marketing strategy." In George Fisk (ed.), Marketing Management Technology as a Social Process, pp. 247-262. New York: Praeger.

Schank, R.C., and R. Abelson. 1977. Scripts, Plans, Goals and Understanding. Hillsdale, NJ: Erlbaum.

Schriber, Jacquelyn B., and Barbara A. Gutek. 1987. "Some time dimensions of work: the measurement of an underlying aspect of organization culture." Journal of Applied Psychology, 72(4):642-650.

Sherwood, Frank P. 1968. "The clock and the specialized R and D society." In Evelyn Glatt and Maynard W. Shelly (eds.), The Research Society, pp. 67-82. New York: Gordon and Breach.

Sherwood, Frank P. 1970.
"Leadership, organizations, and time." In Dwight Waldo, Temporal Dimensions of Development Administration, pp. 216-227. Durham, NC: Duke University Press.

Shimp, Terence A. 1982.
"Perception of time in comsumer research: comment." Journal of Comsumer Research, 9:116-118.

Smith, George David, and Laurence E. Steadman. 1981.
"Present value of corporate history." Harvard Business Review, November-December.

Stalk, Jr., George. 1988.
"Time -- the next source of competitive advantage." Harvard Business Review, July-August:44-51.

Svenson, Ola. 1984.
"Cognitive processess in judging cumulative risk over different periods of time." Organizational Behavior and Human Performance, 33(1):22-41.

Svenson, Ola. 1985.
"Cognitive strategies in a complex judgment task: analysis of concurrent verbal reports and judgments of cumulated risk over different exposure times." Organizational Behavior and Human Decision Processes, 36(1):1-15.

Talbot, F. Brian. 1982.
"Resource-constrained project scheduling with time-resource tradeoffs: the nonpreemptive case." Management Science, 28:1197-1210.

Venkatesan, M., and Johan Arndt. 1980.
"Temporal dimension of consumer behavior: an exploration with time budget." Research in Marketing, 3:163-234.

Vroom, Victor H., and Bernd Pahl. 1971.
"Relationship between age and risk taking among managers." Journal of Applied Psychology, 55:399-405.

Waldo, Dwight. 1970.
Temporal Dimensions of Development Administration. Durham, NC: Duke University Press.

Wright, Peter. 1974.
"The harassed decision maker: time pressures, distractions, and the use of evidence." Journal of Applied Psychology, 59:555-561.

Wright, Peter, and Barton Weitz. 1977.
"Time horizon effects on product evaluation strategies." Journal of Marketing Research, 14(4):429-443.

Zakay, Dan, and Stuart Wooler. 1984.
"Time pressure, training and decision effectiveness." Ergonomics, 27(3):273-284.

13
Music

Beauvillain, C., and P. Fraisse. 1984.
"On the temporal control of polyrhythmic performance." Music Perception, 1(4):485-499.

Becker, Judith. 1981.
"Hindu-Buddhist time in Javanese Gamelan Music." In J.T. Fraser, et al., (eds.) , The Study of Time IV, pp. 161-172. New York: Springer-Verlag.

Bickel, Frank. 1984.
"A time-velocity ration investigation." Journal of Research in Music Education, 32(2): 105-111.

Butler, D., and H. Brown. 1984.
"Tonal structure versus function: study of the recognition of harmonic function." Music Perception, 2:6-24.

Chapple, E.D. 1982.
"Movement and sound: the musical language of body rhythms in interaction." In M. Davis (ed.), Interaction Rhythms: Periodicity in Communication Behavior, pp. 31-51. New York: Human Sciences Press.

Clarke, E. 1982.
"Timing in the performance of Erik Satie's 'Vexations'." Acta Psychologica, 50:1-19.

Deutsch, Diana., and J. Feroe. 1981.
"The internal representation of pitch sequences in tonal music." Psychological Review, 88:503-522.

Durr, Walther. 1981.
"Rhythm in music: a formal scaffolding of time." In J.T. Fraser (ed.), The Voices of Time, 2nd edn., pp. 180-200. Amherst, MA: University of Massachusetts Press.

Epstein, David. 1979.
Beyond Orpheus: Studies in Musical Structure. Cambridge, MA: MIT Press.

Epstein, David. 1990.
The Sounding Stream: Studies in Musical Time. New York: Schirmer Books.

Fraisse, Paul. 1982.
"Rhythm and tempo." In D. Deutsch (ed.), The Psychology of Music, pp. 149-180. New York: Academic.

Handel, S. 1974.
"Perceiving melodic and rhythmic auditory patterns." Journal of Experimental Psychology, 103:922-933.

Handel, S., and J.S. Oshinsky. 1981.
"The meter of syncopated auditory polyrhythms." Perception and Psychophysics, 30:1-9.

Handel, S., M.S. Weaver, and G. Lawson. 1983.
"Effect of rhythmic grouping on stream segregation." Journal of Experimental Psychology: Human Perception and Performance, 9:637-651.

Idson, W.L, and D.W. Massaro. 1976.
"Cross-octave masking of single tones and musical sequences: the effects of structure on auditory recognition." Perception and Psychophysics, 19:155-175.

Kramer, Jonathan D. 1986.
"Temporal linearity and nonlinearity in music." In J.T. Fraser, N. Lawrence, and F.C. Haber (eds.), Time, Science, and Society in China and the West: The Study of Time V, pp. 126-137. Amherst, MA: University of Massachusetts Press.

Kramer, Jonathan D. 1989.
The Time of Music: New Meanings, New Temporalities, New Listening Strategies. New York: Schirmer Books.

Longuet-Higgins, H.C., and C.S. Lee. 1984.
"The rhythmic interpretation of monophonic music." Music Perception, 1:424-441.

Pasler, Jann. 1989.
"Narrative and narrativity in music." In J.T. Fraser (ed.), Time and Mind: The Study of Time VI. Madison, CT: International Universities Press.

Povel, Dirk-Jan. 1977.
"Temporal structure of performed music." Acta Psychologica, 41:309-320.

Povel, Dirk-Jan, and Peter Essens. 1985.
"Perception of temporal patterns." Music Perception, 2(4):411-440.

Rasch, R.A. 1979.
"Synchronization in performed ensemble music." Acoustica, 43:121-131.

Rochberg, G. 1975.
"The structure of time in music: traditional and contemporary ramifications and consequences." In J.T. Fraser and N. Lawerence (eds.), The Study of Time II, pp. 136-149. New York: Springer-Verlag.

Rowell, Lewis. 1978.
"Time in the musical consiousness of old high civilizations: East and West." In J.T. Fraser, N. Lawrence, and D. Park (eds.), The Study of Time III, pp. 578-611. New York: Springer-Verlag.

Rowell, Lewis. 1983.
Thinking About Music: An Introduction to the Philosophy of Music. Amherst, MA: University of Massachusetts Press.

Rowell, Lewis. 1986.
"Thinking time and thinking about time in Indian music." Communication and Cognition, 19(2):229-240.

Shaffer, L. Henry. 1984.
"Timing in musical performance." In John Gibbon and Lorraine Allan (eds.) _Timing and Time Perception_, pp. 420-428. New York: New York Academy of Sciences.

Shaffer, L. Henry, E. Clarke, and N. Todd. 1984.
"Metre and rhythm in piano playing." _Cognition_.

Simon, H.A., and R.K. Summer. 1968.
"Pattern in music." In B. Kleinmuntz (ed.), _Formal Representation of Human Judgment_, pp. 219-250. New York: Wiley.

Sternberg, Saul, and Ronald L. Knoll. 1984.
"Perception, production, and imitation of time ratios by skilled musicians." In John Gibbon and Lorraine Allan (eds.), _Timing and Time Performance_, pp.429-441. New York: New York Academy of Sciences.

Sternberg, Saul, and Ronald L. Knoll, and P. Zukofsky. 1982.
"Timing by skilled musicians." In D. Deutsch (ed.), _The Psychology of Music_, pp. 181-239. New York: Academic Press.

Stone, Ruth M. 1986.
"The shape of time in African music." In J.T. Fraser, N. Lawrence, and F.C. Haber (eds.), _Time, Science, and Society in China and the West: The Study of Time V_, pp. 113-125. Amherst, MA: University of Massachusetts Press.

Terhardt, Ernst. 1984.
"The concept of musical consonance: a link between music and psychoacoustics." _Music Perception_, 1(3):276-295.

Wang, Cecilia C., and Rita S. Salzberg. 1984.
"Discrimination of modulated music tempo by string students." _Journal of Research in Music Education_, 32(2):123-131.

Yates, S. 1935.
"Some aspects of time difficulties and their relation to music." _International Journal of Psycho-Analysis_, 16:341-354.

Yeston, M. 1976.
The Stratification of Musical Rhythm. New Haven, CT: Yale University Press.

14
Philosophy and Physics

Aaronson, Bernard S. 1972.
"Time, time stance, and existence." In J.T. Fraser, F.C. Haber, and G.H. Muller (eds.), The Study of Time, pp. 293-311. New York: Springer-Verlag.

Ackerman, Robert. 1971.
"The fallacy of conjunctive analysis." In Eugene Freeman and Wilfrid Sellars (eds.), Basic Issues in the Philosophy of Time, pp. 154-163. La Salle, IL: Open Court.

Adler, William. 1989.
Time Immemorial: Archaic History and its Sources in Christian Chronography fron Julius Africanus to George Syncellus. Washington, DC: Dumbarton Oaks.

Agazzi, Evandro. 1978.
"The riddle of time." Epistemologia, 1:201-204.

Agazzi, Evandro. 1978.
"Time and causality." Epistemologica, 1:397-424.

Aguessy, Honorat, et al. 1977.
Time and the Philosophies. Paris, France: Unesco Press.

Akhundov, Murad D. 1986.
Conceptions of Space and Time: Sources, Evolution, Directions. Cambridge, MA: MIT Press.

Al-Azm, S.J. 1967.
Kant's Theory of Time. New York: Philosophical Library.

Albritton, Jr., Claude C. 1980.
The Abyss of Time. San Francisco, CA: Freeman, Cooper.

Alexander, Samuel. 1927.
Space, Time, and Deity: The Gifford Lectures at Glasgow, Vol. 1. London, UK: Macmillan.

Anscombe, G.E.M. 1964.
"Before and after." Philosophical Review, 73:3-24.

Anscombe, G.E.M. 1975.
"Causality and determination." In Ernest Sosa (ed.), Causation and Conditionals, pp. 63-81. Oxford, UK: Oxford University Press.

Ariotti, P.E. 1975.
"The concept of time in western antiguity." In J.T. Fraser and N. Lawrence (eds.), The Study of Time II, pp. 69-80. New York: Springer-Verlag.

Aristotle. 1978.
"Time." In Richard M. Gale (ed.), The Philosophy of Time: A Collection of Essays, pp. 9-23. Atlantic Highlands, NJ: Humanities.

Armstrong, David M. 1980.
"Identity through time." In Peter van Inwagen (ed.), Time and Cause: Essays Presented to Richard Taylor, pp. 67-78. Dordrecht, Netherlands: Reidel.

Aronov, R.A. 1971.
"On the foundations of the hypothesis of discrete character of space and time." In Jiri Zeman (ed.), Time in Science and Philosophy, pp. 265-274. New York: Elsevier.

Ashish, Sri Madhava. 1977.
"The Guru as exemplar." In H. Aguessy et al., Time and the Philosophies, pp. 211-30. Paris, France: Unesco Press.

Askin. Yakov F. 1977.
"The philosophic concept of time (a Soviet view)." In H.

Aguessy et al., Time and the Philosophies, pp. 127-39. Paris, France: Unesco Press.

Atkins, P.W. 1987.
"Time and dispersal: the second law." In Raymond Flood and Michael Lockwood (eds.), The Nature of Time, pp. 80-98. New York: Basil Blackwell.

Ayer, A.J. 1954.
"Statements about the past." In A.J. Ayer, Philosophical Essays. London, UK: Macmillan.

Baker, Lynne Rudder. 1974-75.
"Temporal becoming: the argument from physics." Philosophical Forum, 6:218-236.

Baker, Lynne Rudder. 1979.
"On the mind-dependence of temporal becoming." Philosophy and Phenomenological Research, 39:341-357.

Balslev, Anindita Niyogi. 1983.
A Study of Time in Indian Philosophy. Wiesbaden, FRG: Otto Harrassowitz.

Balslev, Anindita Niyogi. 1986.
"Reflections on time in Indian philosophy: with comments on so-called cyclic time." In J.T Fraser, N. Lawrence, and F.C. Haber (eds.), Time, Science, and Society in China and the West: The Study of Time V, pp. 104-112. Amherst, MA: University of Massachusetts Press.

Barnes, J.A. 1971.
"Time flies like an arrow." Man (N.S.):6:537-552.

Barrett, William. 1978.
"The flow of time." In Richard M. Gale (ed.), The Philosophy of Time: A Collection of Essays, pp. 355-377. Atlantic Highlands, NJ: Humanities.

Basri, A.S. 1966.
A Deductive Theory of Space and Time. Amsterdam, Netherlands: North Holland.

Bedini, S.A. 1975.
"Oriental concepts of the measure of time." In J.T. Fraser and N. Lawrence, (eds.) The Study of Time II, pp. 451-484. New York: Springer-Verlag.

Benjamin, A. Cornelius. 1981.
"Ideas of time in the history of philosophy." In J.T. Fraser (ed.), The Voices of Time, 2nd edn., pp. 13-30. Amherst, MA: University of Massachusetts Press.

Bennett, J. 1984.
"Counterfactuals and temporal direction." Philosophical Review, 93:57-91.

Benthem, Johan Van. 1980.
"Points and periods." In Christian Rohner (ed.), Time, Tense, and Quantificaters, pp. 39-57. Tubingen: Neimeyer.

Benthem, Johan Van. 1982.
The Logic of Time: A Model-Theoretic Investigation into the Varieties of Temporal Ontology and Temporal Discourse. (Synthese Library, Vol. 156). Boston, MA: Reidel.

Benthem, Johan Van. 1984.
"Tense logic and time." Notre Dame Journal of Formal Logic, 25:1-16.

Bentley, Arthur F. 1941.
"The factual space and time of behavior." Journal of Philosophy, 38:447-485.

Berger, G.A. 1967.
"Phenomenological approach to the problem of time." In N. Lawrence and D. O'Connor (eds.), Readings in Existential Phenomenology. Englewood Cliffs, NJ: Prentice-Hall.

Berger, George. 1973.
"Temporally symmetric causal relations in Minkowski space-time." In Patrick Supppes (ed.), Space, Time and Geometry, pp. 56-71. Dordrecht, Netherlands: Reidel.

Bergson, Henri. 1889/1960.
Time and Freewill: An Essay on the Immediate Data of Consciousness. New York: Torchbooks.

Bergson, Henri Louis. 1922/1965.
Duration and Simultaneity, with Reference to Einstein's Theory. Indianapolis, IN: Bobbs-Merrill.

Bhattacharya, U.C. 1934.
"Problems of time in Indian Thought." Calcutta Review, 52:302-309.

Bigelow, Patrick. 1986.
"The indeterminability of time in 'Sein and Zeit'." Philosophy and Phenomenological Research, 46:357-379.

Black, Max. 1962.
"The direction of time." In Max Black, Models and Metaphors. Ithaca, NY: Cornell University Press.

Blake, R.M. 1925.
"On Mr. Broad's theory of time." Mind, 34.

Bohm, David. 1971.
Causality and Chance in Modern Physics. Philadelphia, PA: University of Pennsylvania Press.

Bondi, H. 1962.
"Physics and cosmology." The Observatory, 82.

Bossert, Philip J. 1976.
"Hume and Husserl on time and time-consciousness." Journal of the British Society for Phenomenology, 7(1):44-55.

Brabant, F.H. 1937.
Time and Eternity in Christian Thought. London, UK: Longmans Green.

Bradley, R.D. 1959.
"Must the future be what it is going to be?" Mind, 68:193-208.

Brandon, S.G.F. 1951.
Time and Mankind. London, UK: Hutchinson.

Brandon, S.G.F. 1963.
Man and His Destiny in the Great Religions. Manchester, UK: Manchester University Press.

Brandon, S.G.F. 1964.
"Time as god and devil." Bulletin of the John Rylands Libary, 47:12-31.

Brandon, S.G.F. 1965.
History, Time and Deity: A Historical and Comparative Study of the Conception of Time in Religious Thought and Practice. Manchester, UK: Manchester University Press.

Brandon, S.G.F. 1972.
"The deification of time." In J.T. Fraser, F.C. Haber, and G.H. Muller (eds.), The Study of Time, pp. 370-382. New York: Springer-Verlag.

Brandon, S.G.F. 1981.
"Time and the destiny of man." In J. T. Fraser (ed.), The Voices of Time, 2nd edn., pp. 140-157. Amherst, MA: University of Massachusetts Press.

Brillouin, L. 1971.
"The arrow of time." In Jiri Zeman (ed.), Time in Science and Philosophy, pp. 101-110. New York: Elsevier.

Brinkley, Alan B. 1960.
"Time in Hegel's phenomenology." Tulane Studies in Philosophy, 9:3-15.

Broad, C.D. 1922.
"Time." In James Hastings et al. (eds.), Encyclopedia of Religion and Ethics, Vol. 12. New York: Scribner's.

Broad, C.D. 1938.
An Examination of McTaggart's Philosophy. (Vol. II, Chapter 35). Cambridge, UK: Cambridge University Press.

Broad, C.D. 1978.
"Ostensible temporality." In Richard M. Gale (ed.), The Philosophy of Time: A Collection of Essays, pp. 117-142. Atlantic Highlands, NJ: Humanities.

Brockelman, Paul. 1981.
"The awareness of time." Research in Phenomenology, 11:211-232.

Brockelman, Paul. 1985.
Time and Self: Phenomenologial Explorations. New York: Crossroad/Scholars.

Broniak, Christopher J. 1983.
"Kierkegaards's fullness of time." Kinesis, 12 (Spring):81-96.

Brown, Delmer. 1974.
"Buddhism and historical thought in Japan before 1221." Philosophy East and West, 24:215-225.

Brumbaugh, Robert S. 1984.
Unreality and Time. Albany, NY: State University of New York Press.

Buckley, Jerome Hamilton. 1966.
The Triumph of Time. Cambridge, MA: Harvard University Press.

Bullock, Merry, Rochel Gelman, and Renee Baillargeon. 1982.
"The development of casual reasoning." In William J. Friedman (ed.), The Developmental Psychology of Time, pp. 209-254. New York: Academic.

Bunge, Mario. 1959.
Causality: The Place of the Causal Principle in Modern Science. Cambridge, MA: Harvard University Press.

Bunge, Mario. 1968.
"Conjunction, succession, determination, and causation." International Journal of Theoretical Physics, 1:299-315.

Bunge, Mario. 1972.
"Time asymmetry, time reversal, and irreversibility." In J.T. Fraser, F.C. Haber, and G.H. Muller (eds.), The Study of Time, pp. 112-130. New York: Springer-Verlag.

Burgess, John. 1978.
"The unreal future." Theoria, 44(3):157-174.

Burgess, John. 1979.
"Logic and time." Journal of Symbolic Logic, 44:566-582.

Burgess, John. 1980.
"Decidability for branching time." Studia Logica, 39:203-218.

Burks, A.W. 1951.
"The logic of casual propositions." Mind, 60:363-382.

Burton, Robert G. 1976.
"The human awareness of time: an analysis." Philosophy and Phenomenological Research, 36:303-318.

Butler, R.J. 1955.
"Aristotle's sea-flight and three-valued logic." Philosophical Review, 64.

Butterfield, J. 1984.
"Seeing the present." Mind, 93:161-176.

Butterfield, J. 1985.
"Spatial and temporal parts." Philosophical Quarterly, 35:32-44.

Cahn, Steven M. 1967.
Fate, Logic and Time. New Haven, CT: Yale University Press.

Cahn, Steven M. 1964.
"Fatalistic arguments." Journal of Philosophy, 61:295-305.

Caldirola, Piero, and Erasmo Recami. 1978.
"The concept of time in physics." Epistemologia, 1:263-304.

Callahan, John F. 1968.
Four Views of Time in Ancient Philosophy. New York: Greenwood.

Capek, Milic. 1960.
"The theory of eternal recurrence in modern philosophy of science, with special reference to C.S. Peirce." Journal of Philosophy, 57:289-296.

Capek, Milic. 1972.
"The friction of instants." In J.T. Fraser, F.C. Haber, and G.H. Muller (eds.), The Study Time, pp. 332-344. New York: Springer-Verlag.

Capek, Milic (ed.). 1976.
The Concepts of Space and Time: Their Structure and Their Development. (Vol. XXII of the Boston Studies in the Philosophy of Science). Boston, MA: Reidel.

Capek, Milic. 1981.
"Time in relativity theory: arguments for a philosophy of becoming." In J. T. Fraser (ed.), The Voices of Time, 2nd edn., pp. 434-454. Amherst, MA: University of Massachusetts Press.

Carr, David. 1986.
Time, Narrative, and History. Bloomington,IN: Indiana University Press.

Carrier, L.S. 1969.
"The time-gap argument." Australasian Journal of Philosophy, 47:263-272.

Carrier, L.S. 1972.
"Time-gap miopia." Austrlasian Journal of Philosophy, 50:55-57.

Carroll, J.A. 1959.
"An absolute scale of time." Nature, 84:260.

Cartwright, N. 1979.
"Causal laws and effective strategies." Nous, 13:419-437.

Casey, Edward S. 1983.
"Keeping the past in mind." Review of Metaphysics, 37:77-95.

Cassirer, Ernst. 1957.
Philosophy of Symbolic Forms, Vol. III: The Phenomenology of Knowledge. New Haven, CT: Yale University Press.

Cassirer, Eva. 1972.
"On the reality of becoming." In J.T. Fraser, F.C. Haber, and G.H. Muller (eds.), The Study of Time, pp. 345-353. New York: Springer-Verlag.

Chacalos, Elias Harry. 1989.
Time and Change: Short But Different Philosophies. Rockville, MD: Potomac Press Circle.

Chapman, T. 1982.
Time: A Philosophical Analysis. Boston, MA: Reidel.

Chappell, V.C. 1962.
"Time and Zeno's arrow." Journal of Philosophy, 59.

Chari, C.T.K. 1958.
"A note on multidimensional time." British Journal for the Philosophy of Science, 8:155-158.

Chari, C.T.K. 1971.
"Informo-dynamics and the anisotropy of time." In Jiri Zeman (ed.), Time in Science and Philosophy, pp. 209-225. New York: Elsevier.

Charlton, William. 1981.
"Time." Philosophy, 56:149-160.

Cheng, C.C. 1974.
The Buddhist Teaching of Totality. University Park, PA: Pennsylvania State University Press.

Cheng, Chung-Ying. 1974.
"Greek and Chinese views on time and the timeless." Philosophy East and West, 24:155-159.

Cheng, Chung-Ying. 1976.
"Model of causality in Chinese philosophy: a comparative study." Philosophy East and West, 26(1):13.

Cheng, Chung-Ying. 1983.
"On the hierarchical theory of time: with reference to Chinese philosophy." Journal of Chinese Philosophy, 10:357-383.

Chessnick, Richard D. 1957.
"The sense of reality, time, and creative inspiration." American Imago, 14:317-331.

Chisholm, Roderick M. 1980.
"Beginning and endings." In Peter van Inwagen (ed.), Time and Cause: Essays Presented to Richard Taylor, pp. 17-25. Dordrecht, Netherlands: Reidel.

Chisholm, Roderick M. 1981.
"Brentano's analysis of the consciousness of time". Midwestern Studies in Philosophy, 6:3-16.

Chisholm, Roderick M., and Richard Taylor. 1960.
"Making things to have happened." Analysis, 20:73-78.

Christensen, Ferrel. 1976.
"The source of the river of time." Ratio, 18:131-144.

Christensen, Ferrel. 1978.
"The theory of space-like time." In J.T. Fraser and D.

Parks (eds.), The Study of Time III, pp. 167-184. New York: Springer-Verlag.

Christensen, Ferrel. 1980. "Two concepts of time." Journal of Chinese Philosophy, 7:327-339.

Clarke, Bowman L. 1983. "Process, time and God." Process Studies, 13:245-259.

Clemence, G.M. 1952. "Time and its measurement." American Scientist, 40:260-269.

Cleugh, Mary F. 1970. Time and its Importance in Modern Thought. New York: Russell and Russell.

Coddington, A. 1975. "Creaking semaphore and beyond: a consideration of Shackle's 'Epistemics and Economics'." British Journal for the Philosophy of Science, 26:151-163.

Coe, Lee. 1969. "The nature of time." American Journal of Physics, 37:810-815.

Collingwood, Robin George. 1975. "Some perplexities about time." In Charles M. Sherover, The Human Experience of Time, pp. 558-571. New York: New York University Press.

Coomaraswamy, A.K. 1947. Time and Eternity. Ascona, Switzerland: Arbus Asiae Publisher.

Corish, Denis. 1978. "Aristotle on temporal order: 'now,' before,' and 'after.' Isis, 69:70-75.

Corish, Denis. 1978. "Time, space and freewill: the Leibnitz-Clarke correspondence." In J.T. Fraser, N. Lawrence, and D. Park (eds.), The Study of Time III, pp. 634-655. New York: Springer-Verlag.

Corish, Denis. 1981.
"The beginning of the beginning in Western thought." In J.T. Fraser et al. (eds.), <u>The Study of Time IV</u>, pp 34-45. New York: Springer-Verlag.

Corish, Denis. 1986.
"The emergence of time: a study in the origins of western thought." In J.T. Fraser, N. Lawrence, and F.C. Haber (eds.), <u>Time, Science, and Society in China and the West: The Study of Time V</u>, pp. 69-78. Amherst MA: University of Massachusetts Press.

Costa de Beauregard, Olivier. 1967.
"Two principles of the science of time." <u>Annals of the New York Academy of Sciences</u>, 138:407-421.

Costa de Beauregard, Olivier. 1971.
"Information and irreversibility problems." In Jiri Zeman (ed.), <u>Time in Science and Philosophy</u>, pp. 11-25. New York: Elsevier.

Costa de Beauregard, Olivier. 1972.
"No paradox in the theory of time anisotropy." In J.T. Fraser, F.C. Haber, and G.H. Muller (eds.), <u>The Study of Time</u>, pp. 131-139. New York: Springer-Verlag.

Costa de Beauregard, Olivier. 1978.
"The third storm of the twentieth century: the Einstein paradox." In J.T. Fraser, N. Lawrence and D. Park (eds.), <u>The Study of Time III</u>, pp 53-73. New York: Springer-Verlag.

Costa de Beauregard, Olivier. 1979.
"Two lectures on the direction of time." In Wesley C. Salmon (ed.), <u>Hans Reichenbach: Logical Empiricist</u>, pp. 341-366. Dordrecht, Netherlands: Reidel.

Costa de Beauregard, Olivier. 1981.
"Time in relativity theory: arguments for a philosophy of being." In J.T. Fraser (ed.), <u>The Voices of Time</u>, 2nd edn., pp. 417-433. Amherst, MA: University of Massachusetts Press.

Costa de Beauregard, Olivier. 1987.
<u>Time, the Physical Magnitude</u>. Dordrecht, Netherlands: Reidel.

Craig, William Lane. 1979.
"Whitrow and Popper on the impossibility of an infinite past." British Journal for the Philosophy of Science, 30:165-170.

Craig, William Lane. 1985.
"Was Thomas Aquinas a B-theorist of time?" The New Scholasticism, 59:475-483.

Cullman, Ocsar. 1950.
Christ and Time: The Primitive Christian Concept of Time and History. Philadelphia, PA: Westminster.

Daniels, Charles B. 1970.
"Seeing through a time-gap." Australasian Journal of Philosophy, 48:354-359.

Das, S.R. 1933.
"Notion of time in Indian philosophy." Indian Historical Quarterly, 9:149-153.

Dauer, D.W. 1975.
"Nietzsche and the concept of time." In J.T. Fraser and N. Lawrence (eds.), The Study of Time II, pp. 81-97. New York: Springer-Verlag.

Davidson, Donald. 1963.
"Actions, reasons, and causes." Journal of Philosophy, 60:685-700.

Davidson, Donald. 1967.
"Causal relations." Journal of Philosophy, 64:691-703.

Davidson, Donald. 1967.
"Truth and meaning." Synthese, 17:304-323.

Davies, P.C.W. 1977.
The Physics of Time Asymmetry. Berkeley, CA: University of California Press.

Davies, P.C.W. 1977.
Space and Time in the Modern Universe. Cambridge, UK: Cambridge University Press.

Davies, P.C.W. 1978.
"Space-time singularities in cosmology and black hole evaporations." In J.T. Fraser, N. Lawrence, and D. Park

(eds.), The Study of Time III, pp. 74-91. New York: Springer-Verlag.

Davies, P.C.W. 1981.
"Time and reality." In R. Healy (ed.) Reduction, Time, and Reality: Studies in the Philosophy of the Natural Sciences, pp. 63-78. Cambridge, UK: Cambridge University Press.

Davies, P.C.W. 1987.
"Time asymmetry and quantum mechanics." In Raymond Flood and Michael Lockwood (eds.), The Nature of Time, pp. 99-124. New York: Basil Blackwell.

De Goeje, C.H. 1949.
What is Time?. Leiden, Netherlands: E.J. Brill.

Denbigh, Kenneth G. 1972.
"In defence of the direction of time." In J.T. Fraser, F.C. Haber, and G.H. Muller (eds.), The Study of Time, pp. 148-158. New York: Springer-Verlag.

Denbigh, Kenneth G. 1978.
"The objectivity, or otherwise, of the present." In J.T. Fraser, N. Lawrence, and D. Park (eds.), The Study of Time III, pp. 307-329. New York: Springer-Verlag.

Denbigh, Kenneth G. 1981.
Three Concepts of Time. New York: Springer-Verlag.

Denbigh, Kenneth G. 1989.
"Physical time and mental time." In J.T. Fraser (ed.), Time and Mind: The Study of Time VI. Madison, CT: International Universities Press.

Deppert, Wolfgang. 1978.
"Remarks on a set theory extension of the concept of time." Epistemologia, 1:425-434.

Deutsh, Eliot. 1982.
Personhood, Creativity and Freedom. Honolulu, HA: University of Hawaii Press.

Devasenapathi, V.A. 1962.
Towards Conquest of Time. Madras, India: University of Madras.

Dewey, John. 1926.
"Events and the future." Journal of Philosophy, 23:253-258.

Dewey, John. 1975.
"Time and individuality." In Charles M. Sherover, The Human Experience of Time, pp. 419-436. New York: New York University Press.

Dick, Philip K. 1967.
Counter-Clock World. New York: Berkley Medallion.

Dingle, Herbert. 1981.
"Time in relativity theory: measurement or coordinate?" In J.T. Fraser (ed.), The Voices of Time, 2nd edn., pp. 455-472. Amherst, MA: University of Massachusetts Press.

Dobbs, H.A.C. 1951.
"The relation between the time of psychology and the time of physics - Part I." British Journal for the Philosophy of Science, 2:122-141.

Dobbs, H.A.C. 1951.
"The relation between the time of psychology and the time of physics - Part II." British Journal for the Philosophy of Science, 2:177-192.

Dobbs, H.A.C. 1972.
"The dimensions of the sensible present." In J.T. Fraser, F.C. Haber, and G.H. Muller (eds.), The Study of Time, pp. 274-292. New York: Springer-Verlag.

Domotor, Zoltan. 1973.
"Causal models and space-time geometries." In Patrick Suppes (ed.), Space, Time and Geometry, pp. 3-55. Dordrecht, Netherlands: Reidel.

Doob, Leonard. 1968.
The Plans of Men. Hamden, CT: Archon Books.

Doss, S.R. 1970.
"Copernicus revisited: time versus 'time' versus time." Philosophy and Phenomenological Research, 31:193-211.

Dretske, F.I. 1962.
"Moving background in time." Philosophical Review, 71:94-98.

Dreyfus, H.L. 1975.
"Human temporality." In J.T. Fraser and N. Lawrence (eds.), The Study of Time II, pp. 150-162. New York: Springer-Verlag.

Ducasse, C.J. 1926.
"On the nature and the observability of the causal relation." Journal of Philosophy, 23:57-68.

Dummett, Michael. 1960.
"A defense of McTaggart's proof of the unreality of time." Philosophical Review, 69:497-504.

Dummett, Michael. 1964.
"Bringing about the past." Philosophical Review, 73:338-359.

Dummett, Michael. 1978.
Truth and Other Enigmas. London, UK: Duckworth.

Dummett, Michael. 1978.
"Can an effect precede its cause?" In Michael Dummett, Truth and Other Enigmas, pp. 319-332. London, UK: Duckworth.

Dummett, Michael. 1978.
"The reality of the past." In Michael Dummett, Truth and Other Enigmas, pp. 358-374. London, UK: Duckworth.

Dummett, Michael. 1987.
"Causal loops." In Raymond Flood and Michael Lockwood (eds.), The Nature of Time, pp. 135-169. New York: Basil Blackwell.

Dunne, J.W. 1959.
An Experiment with Time. Atlantic Highlands, NJ: Humanities.

Dupre, John. 1984.
"Probabilistic causality emancipated." In Peter A. French, Theodore E. Uehling, Jr., and Howard K. Wettstein (eds.), Causation and Causal Theories, pp. 169-175. Minneapolis, MN: University of Minnesota Press.

Earman, John. 1967.
"On going backwards in time." Philosophy of Science, 34:211-222.

Earman, John. 1969.
"The anisotropy of time." Australasian Journal of Philosophy, 47:273-295.

Earman, John. 1970.
"Space-time, or how to solve philosophical problems and dissolve philosophical muddles without really trying." Journal of Philosophy, 67:259-277.

Earman, John. 1970.
"Who's afraid of absolute space?" Australasian Journal of Philosophy, 48:287-319.

Earman, John. 1973.
"Notes on the causal theory of time." In Patrick Suppes (ed.), Space, Time and Geometry, pp. 72-84. Dordrecht, Netherlands: Reidel.

Earman, John. 1974.
"An attempt to add a little direction to 'the problem of the direction of time'." Philosophy of Science, 41:15-47.

Earman, John. 1976.
"Causation: a matter of life and death." Journal of Philosophy, 73:12.

Earman, John. 1977.
"How to talk about the topology of time." Nous, 11(3):211-226.

Earman, John, S., Clark Glymour, and John J. Stachel (eds.). 1977.
Foundations of Space-Time Theories. (Minnesota Studies in the Philosophy of Science, Vol. VIII). Minneapolis, MN: University of Minnesota Press.

Eddington, Arthur S. 1959.
Space, Time and Gravitation. New York: Harper.

Eells, Ellery, and Elliott Sober. 1983.
"Probabilistic causality and the question of transitivity." Philosophy of Science, 50:35-57.

Ehrenfels, Christian. 1948.
Cosmogony. New York: Philosophical Library.

Eisley, Loren. 1980.
The Firmament of Time. New York: Atheneum.

Ekeland, Ivan. 1988.
Mathematics and the Unexpected. Chicago, IL: University of Chicago Press.

Eliade, Mircea. 1958.
"Time and eternity in Indian thought." In Man and Time, Eranos Year Book 1958, pp. 173-200. London, UK: Routledge and Kegan Paul.

Ellis, Brian, and Peter Bowman. 1967.
"Conventionality in distant simultaneity." Philosophy of Science, 34:116-136.

Erickson, Stephen A. 1984.
Human Presence: At the Boundaries of Meaning. Macon, GA: Mercer University Press.

Essen, L. 1973.
The Measurement of Frequency and Time Interval. London, UK: Her Majesty's Stationary Office.

Fagg, Lawrence W. 1985.
Two Faces of Time. Wheaton, IL: Theosophical Society.

Feather, N. 1961.
Mass, Length and Time. Harmondsworth, UK: Penguin.

Ferre, Frederick. 1972.
"Grunbaum on temporal becoming: a critique." International Philosophical Quarterly, 12:426-445.

Feuerbach, Ludwig. 1966.
Principles of the Philosophy of Future. Indianapolis, IN: Bobbs-Merrill.

Field, Richard W. 1983.
"William James and the epochal theory of time." Process Studies, 13:260-274.

Fields, H. 1979.
"On the status of 'the direction of time'." Methodology and Science, 12:213-235.

Findlay, J.N. 1956.
"An examination of tenses." In H.D. Lewis (ed.), Contemporary British Philosophy, Third Series. New York: Macmillan.

Findlay, J.N. 1978.
"Time: a treatment of some puzzles." In Richard M. Gale (ed.), The Philosophy of Time: A Collection of Essays, pp. 143-162. Atlantic Highlands, NJ: Humanities.

Finkelstein, D. 1978.
"Beneath time: exploration in quantum topology." In J.T. Fraser, N. Lawrence, and D. Park (eds.), The Study of Time III, pp. 94-112. New York: Springer-Verlag.

Fisher, John. 1975.
"Fallability and knowledge of the future." Philosophy and Phenomenological Research, 36(1):44-58.

Fitzgerald, Paul. 1969.
"The truth of tomorrow's sea fight." Journal of Philosophy, 66:307-329.

Fitzgerald, Paul. 1974.
"Critical study of 'The Language of Time' by Richard M. Gale." The Philosophical Forum, 5:424-440.

Fitzgerald, Paul. 1985.
"Stump and Kretzmann on time and eternity." Journal of Philosophy, 82:260-269.

Flew, A.G.N. 1959.
"Hobbes and the sea flight." Graduate Review of Philosophy, 2.

Flood, Raymond, and Michael Lockwood (eds.). 1987.
The Nature of Time. New York: Basil Blackwell.

Fokker, A.D. 1965.
Time and Space, Weight and Inertia. Oxford, UK: Pergamon.

Foures, Y., and I.E. Segal. 1955.
"Causality and analyticity." Transactions of the American Mathematical Society, 78:385-405.

Fraser, J.T. 1972.
"The concept of time in Western thought." Main Currents in Modern Thought, 28:115-124.

Fraser, J.T. 1980.
"Out of Plato's cave: the natural history of time." Kenyon Review, 2 (n.s):143-162.

Fraser, J.T. 1982.
The Genesis and Evolution of Time: A Critique of Interpretation in Physics. Amherst, MA: University of Massachusetts Press.

Fraser, J.T. 1989.
"The many dimensions of time and mind: an epistemic jigsaw puzzle game." In J.T. Fraser (ed.), Time and Mind: The Study of Time VI. Madison, CT: International Universities Press.

Fraser, J.T. 1989.
"The elusiveness of the mind." In J.T. Fraser (ed.), Time and Mind: The Study of Time VI. Madison, CT: International Universities Press.

Freedman, D.Z., and P. van Nieuwenhuizen. 1985.
"The hidden dimension of spacetime." Scientific American, 252(3):74-83.

Freeman, Eugene, and Wilfrid Sellars (eds.). 1971.
Basic Issues in the Philosophy of Time. La Salle, IL: Open Court.

Friedman, M. 1983.
Foundations of Space-Time Theories. Princeton, NJ: Princeton University Press.

Frings, Manfred S. 1983.
"Social temporality in George Herbert Mead and Sheler." Philosophy Today, 27, No. 414, pp. 281-89.

Fulmer, Gilbert. 1980.
"Understanding time travel." Southwestern Journal of Philosophy, 11:151-156.

Gale, Richard M. 1962.
"Tensed statements." Philosophical Quarterly, 12.

Gale, Richard M. 1968.
The Language of Time. London, UK: Routledge and Kegan Paul.

Gale, Richard M. 1971.
"'Here' and 'now'." In Eugene Freeman and Wilfrid Sellars (eds.), Basic Issues in the Philosophy of Time, pp. 72-85. La Salle, IL: Open Court.

Gale, Richard M. 1977.
"A reply to Oaklander." Philosophy and Phenomelogical Research, 38:234-238.

Gale, Richard M. 1978.
The Philosophy of Time: A Collection of Essays. Atlantic Highlands, NJ: Humanities.

Galtung, Johan. 1970.
"Diachronic correlation, process analysis and causal analysis: the quest for a diachronic, nomothetic social science." Quantity and Quality, 4:55-94.

Gander, J.E. 1967.
"Some possible interrelationships between steady-state kinetics and time." Annals of the New York Academy of Sciences, 138:730-747.

Gardet, Louis. 1977.
"The prophet." In Louis Gardet et al., Time and the Philosophies, pp. 197-210. Paris, France: Unesco Press.

Gardner, Martin. 1967.
"Can time go backward?" Scientific American, Vol. 216, No. 1 (January), pp. 98-108.

Gardner, Martin. 1979.
"On altering the past, delaying the future and other ways of tampering with time." Scientific American, 240:21-30.

Garson, James W. 1971.
"Here and now." In Eugene Freeman and Wilfrid Sellars (eds.), Basic Issues in the Philosophy of Time, pp. 145-153. La Salle, IL: Open Court.

Geach, P.T. 1965.
"Some problems about time." Proceedings of the British

Academy, Vol. LI, pp. 321-336. London, UK: Oxford University Press.

Gerathewohl, Siegfried. 1967.
"Orientation and navigation in space-time." Annals of the New York Academy of Sciences, 138:504-514.

Gibson, J.J. 1975.
"Events are perceivable but time is not." In J.T. Fraser and N. Lawrence (eds.), The Study of Time II, pp. 295-301. New York: Springer-Verlag.

Gioscia, Victor. 1974.
Time Forms: Beyond Yesterday and Tomorrow. New York: Gordon and Breach.

Glymour, Clark. 1982.
"Causal inference and causal explanation." In Robert McLaughlin (ed.), What? Where? When? Why?: Essays on Induction, Space and Time, Explanation, pp. 179-191. Dordrecht, Netherlands: Reidel.

Gold, Thomas. 1962.
"The arrow of time." American Journal of Physics, 30:403-410.

Gold, Thomas (ed.). 1967.
The Nature of Time. Ithaca, NY: Cornell University Press.

Goldman, Steven L. 1981.
"On the beginnings and endings of time in medieval Judaism and Islam." In J.T. Fraser et al. (eds.), The Study of Time IV, pp. 59-72. New York: Springer-Verlag.

Goldman, Steven L. 1982.
"Modern science and western culture: the issue of time," Antioch Review, 24:371-401.

Gonseth, Ferdinand. 1972.
Time and Method: An Essay on the Methodology of Research. Springfield, IL: Charles C. Thomas.

Good, I.J. 1961.
"A causal calculus, I." British Journal for the Philosophy of Science, 11:305-318.

Good, I.J. 1962.
"A causal calculus, II." British Journal for the Philosophy of Science, 12:43-51.

Goodman, N. 1947.
:The problem of counterfactual conditions." Journal of Philosophy, 44:113-128.

Gorovitz, Samuel. 1964.
"Learning the past alone." Philosophical Review, 73:360-371.

Grunbaum, Adolf. 1962.
"The nature of time." In R.G. Colodny (ed.), Frontiers of Science and Philosophy. Pittsburgh, PA: University of Pittsburgh Press.

Grunbaum, Adolf. 1962.
"Temporally asymmetric principles, parity between explanation and prediction and mechanism versus teleology." Philosophy of Science, 29:146.

Grunbaum, Adolf. 1964.
"Popper on irreversibility." In Adolf Grunbaum (ed.), The Critical Approach: Essays in Honor of Karl Popper. New York: Free Press of Glencoe.

Grunbaum, Adolf. 1964.
"Time, irreversible processes, and the physical status of becoming." In J.J.C. Smart (ed.), Problems of Space and Time, pp. 397-425. New York: Macmillan.

Grunbaum, Adolph. 1967.
"The status of temporal becoming." Annals of the New York Academy of Sciences, 138:374-395.

Grunbaum, Adolf, et al. 1969.
"A panel discussion of slow clock transport in the special and general theories of relativity." Philosophy of Science, 36:1-81.

Grunbaum, Adolph. 1970.
"Space, time and falsifiability, Part I." Philosophy of Science, 37:469-588.

Grunbaum, Adolf. 1971.
"The meaning of time." In Eugene Freeman and Wilfrid

Sellars (eds.), Basic Issues in the Philosophy of Time, pp. 195-228. La Salle: IL: Open Court.

Grunbaum, Adolf. 1973.
Philosophical Problems of Space and Time, 2nd, revised, enlarged edn.). Dordrecht, Netherlands: Reidel.

Grunbaum. Adolf. 1977.
"Absolute and relational theories of space and space-time." In John S. Earman, Clark N. Glymour, and John J. Stachel (eds.), Foundations of Space-Time Theories. Minneapolis, MN: University of Minnesota Press.

Grunbaum, Adolf. 1978.
"The status of temporal becoming." In Richard M. Gale (ed.), The Philosophy of Time: A Collection of Essays, pp. 322-354. Atlantic Highlands, NJ: Humanities.

Gunn, J.A. 1929.
The Problem of Time. London, UK: Allen and Unwin.

Gunter, Pete Addison Y. (ed.). 1969.
Bergson and the Evolution of Physics. Knoxville, TN: University of Tennessee Press.

Gunther, Gotthard. 1967.
"Time, timeless logic and self-referential systems." Annals of the New York Academy of Sciences, 138:396-406.

Hacker, P.M.S. 1982.
"Events and objects in space and time." Mind, 91:1-19.

Hall, David L. 1981.
"From otherness to emptiness: the aesthetics of philosophic communication." Journal of Chinese Philosophy, 8:497-513.

Hamblin, C.L. 1971.
"Starting and stopping." In Eugene Freeman and Wilfrid Sellars (eds.), Basic Issues in the Philosophy of Time, pp. 86-101. La Salle, IL: Open Court.

Hamblin, C.L. 1972.
"Instants and intervals." In J.T. Fraser, F.C. Haber, and G.H. Muller (eds.), The Study of Time, pp. 324-331. New York: Springer-Verlag.

Hammerschmidt, William W. 1975.
Whitehead's Philosophy of Time. New York: Russell and Russell.

Hankins, Thomas L. 1976.
"Algerbra as pure time: William Rowan Hamilton and the foundations of algebra." In Peter K. Machamer and Robert G. Turnbull (eds.), Motion and Time, Space and Matter, pp. 326-359. Columbus, OH: Ohio State University Press.

Hardy, A., R. Harvie, and A. Koestler. 1974.
The Challenge of Chance. New York: Random House.

Harris, Errol E. 1976.
"Time and eternity." The Review of Metaphysics, 29:464-482.

Harris, Errol E. 1988.
The Reality of Time. Albany, NY: State University of New York Press.

Harrison Craig. 1973.
"On the structure of space-time." In Patrick Suppes (ed.), Space, Time and Geometry, pp. 178-192. Dordrecht, Netherlands: Reidel.

Hasker, William. 1989.
God, Time and Knowledge. Ithaca, NY: Cornell University Press.

Hasnaoui, Ahmed. 1977.
"Certain notions of time in Arab-Muslim philosophy." In Louis Gardet et al., Time and the Philosophies, pp. 49-79. Paris, France: Unesco Press.

Hawking, S.W. 1968.
"The existence of cosmic time functions." Proceedings of the Royal Society (London), A 308:433-435.

Hawking, S.W., and G.F.R. Ellis. 1973.
The Large Scale Structure of Space-Time. Cambridge, UK: Cambridge University Press.

Hawking, Stephen M. 1988.
A Brief History of Time: From the Big Bang to Black Holes. New York: Bantam.

Healy, R. (ed.). 1981.
Reduction, Time and Reality: Studies in the Philosophy of the Natural Sciences. Cambridge, UK: Cambridge University Press.

Heath, L.R. 1956.
The Concept of Time. Chicago, IL: University of Chicago Press.

Hegel, G.W.F. 1975.
"Time and becoming." In Charles M. Sherover, The Human Experience of Time, pp. 183-192. New York: New York University Press.

Heidegger, Martin. 1962.
Being and Time. New York: Harper & Row.

Heidegger, Martin. 1979.
"The understanding of time in phenomenology and in the thinking of the being-question." Southwestern Journal of Philosophy, 10(2):199-201.

Heidegger, Martin. 1985(1925).
History of the Concept of Time: Prolegomena. Bloomington, IN: Indiana University Press.

Heine, Steven. 1985.
Existential and Ontological Dimensions of Time in Heidegger and Dogen. Albany, NY: State University of New York Press.

Heinz, Marion. 1986.
"The concept of time in Heidegger's early works." In Joseph J. Kockelmans (eds.), A Companion to Martin Heidegger's 'Being and Time', pp. 183-207. Washington, DC: Center for Advanced Research in Phenomenology and University Press of America.

Helm, Bertrand P. 1975.
"William James on the nature of time." Tulane Studies in Philosophy, 24:33-47.

Helm, Bertrand P. 1980.
"The nature and modes of time." Monist, 63:375-385.

Helm, Bertrand P. 1985.
Time and Reality in American Philosophy. Amherst, MA: University of Massachusetts Press.

Herbert, G.S. 1978.
Time: A Metaphysical Study. Trivandrum, India: College Book House.

Hering, D.W. 1962.
"The concept of time sense among cultured and uncultured peoples." In H. Shapely (ed.), Time and Its Mysteries, pp. 83-111. New York: Collier.

Hesslow, G. 1981.
"Causality and determinism." Philosophy of Science, 48:591-605.

Highwater, Jamake. 1981.
"Time." In The Primal Mind, pp. 89-118. New York: Harper and Row.

Hinckfuss, Ian. 1975.
The Existence of Space and Time. Oxford, UK: Clarendon Press.

Hintikka, Jaakko. 1973.
Time and Necessity: Studies in Aristotle's Theory of Modality. New York: Oxford University Press.

Hinton, C.H. 1887.
What is the Fourth Dimension? London, UK: Allen and Unwin.

Hoffman, Piotr. 1973.
"A note on time and subjectivity in Kant's Critique of Pure Reason." Philosophical Forum, 4:313-322.

Hoffman, Piotr. 1987.
Doubt, Time, Violence. Chicago, IL: University of Chicago Press.

Holubar, J. 1969.
The Sense of Time. Cambridge, MA: MIT Press.

Horwich, Paul. 1987.
Asymmetries in Time: Problems in the Philosophy of Science. Cambridge, MA: MIT Press.

Houts, Ronald W. 1980.
"Some implications of the time-lag argument." Philosophy and Phenomenological Research, XLI(1-2):150-157.

Hoy, David Couzens. 1978.
"History, historicity and historiography in Being and Time." In Michael Murray (ed.), Heidegger and Modern Philosophy, pp. 339-340. New Haven, CT: Yale University Press.

Hoyt, Robert S., and Leonid Hurwicz (eds.). 1966.
Man vs Time. Minneapolis, Minnesota: University of Minnesota Press.

Hsu, J.P. 1979.
"The analysis of time: is the relativistic time unique?" Foundations of Physics, 9:55-70.

Huertas-Jourda, J. 1975.
"Structures of the 'living-present': Husserl and Proust." In J.T. Fraser and N. Lawrence (ed.), The Study of Time II, pp. 163-195. New York: Springer-Verlag.

Huisheng, Lo. 1986.
"Zi Wu flow theory and time." In J.T. Fraser, N. Lawrence, and F.C. Haber (eds.), Time, Science, and Society in China and the West: The Study of Time V, pp. 219-224. Amherst, MA: University of Massachusetts Press.

Humberstone, I. 1979.
"Internal semantics for tense logics." Journal of Philosophical Logic, 8:171-196.

Husserl, Edmund. 1964.
The Phenomenology of Internal Time-Consciousness. Bloomington, IN: Indiana University Press.

Inada, Kenneth K. 1974.
"Time and temporality -- a Buddhist Approach." Philosophy East and West, 24:171-179.

Ingarden, Roman. 1964.
Time and Modes of Being. Springfield, IL: Thomas.

James, William. 1975.
"The perception of time." In Charles M. Sherover, The

Human Experience of Time, pp. 368-383. New York: New York University Press.

Janich, Peter. 1985.
Protophysics of Time: Constructive Foundation and History of Time Management. Dordnecht, Netherlands: Reidel.

Jaques, Elliott. 1982.
The Form of Time. New York: Crane, Russak.

Jobe, Evan, K. 1976.
"Temporal predication and identity." Australasian Journal of Philosophy, 54:65-71.

Jobe, Evan K. 1980.
"Nature's choice of time." Australasian Journal of Philosophy, 58:347-359.

Johnson, Conrad Dale. 1986.
"Scientific explanation of the evolution of time." In J.T. Fraser, N. Lawrence, and F.C. Haber (eds.), Time, Science, and Society in China and the West: The Study of Time V, pp. 39-50. Amherst, MA: University of Massachusetts Press.

Jones, M.R. 1981.
"Only time can tell: on the topology of mental space and time." Critical Inquiry, 7:557-576.

Jordan, Richard Douglas. 1972.
The Temple of Eternity; Thomas Traherne's Philosophy of Time. Port Washington, NY: Kemikat Press.

Jourdain, P.E.B. 1916.
"The flying arrow." Mind, 25.

Jung, Carl G. 1960.
"Synchronicity: an acausal connecting principle." In Works of C.G. Jung, Vol. 8, pp. 417-531. London, UK: Routledge and Kegan Paul.

Jung, Carl G. 1973.
Synchronicity. Princeton, NJ: Princeton University Press.

Kalupahana, David J. 1974.
"The Buddhist conception of time and temporality." Philosophy East and West, 24:181-191.

Kalupahana, David J. 1975.
Causality: The Central Philosophy of Buddhism. Honolulu, HA: University Press of Hawaii.

Kant, Immanuel. 1975.
"The primacy of time." In Charles M. Sherover, The Human Experience of Time, pp. 143-152. New York: New York University Press.

Kant, Immanuel. 1975.
"The temporalization of concepts." In Charles M. Sherover, The Human Experience of Time, pp. 153-156. New York: New York University Press.

Kaplan A. 1965.
"Non-causal explanation." In D. Lerner (ed.), Cause and Effect, pp. 145-155. New York: Free Press.

Kelly, George A. 1958.
"Man's construction of his alternatives." In Gardner Lindzey (ed.), Assessment of Human Motives, pp. 33-64. New York: Rinehart and Co.

Khatchadourian, Haig. 1973.
"Do ordinary spatial and temporal expressions designate relations?" Philosophy and Phenomenological Research, 34:82-94.

Khundmiri, S. Alam. 1974.
"Eastern Aristotelians and time." Indian Journal of Philosophic Studies, 1(1):19-23.

Kim, Jaegwon. 1984.
"Epiphenomenal and supervenient causation." In Peter A. French, Theodore E. Uehling, Jr., and Howard K. Wettstein (eds.), Causation and Causal Theories, pp. 257-281. Minneapolis, MN: University of Minnesota Press.

Kline, George L. 1986.
"'Present', 'past', and 'future' as categorical terms, and the 'falacy of the actual future'." Review of Metaphysics, 40:215-235.

Kneale, W.C. 1960-61.
"Time and eternity in theology." Proceedings of the Aristotelian Society, 61(n.s.):87-108.

Kockelmans, Joseph J. (ed.). 1986.
A Companion to Martin Heidegger's "Being and Time". Washington, DC: Center for Advanced Research in Phenomenology and University Press of America.

Kolb, David A. 1974.
"Time and the timeless in Greek thought." Philosophy East and West, 24:137-143.

Koller, John M. 1974.
"On Buddhist views of devouring time." Philosophy East and West, 24:201-208.

Koslow, Arnold (ed.). 1967.
The Changeless Order: The Physics of Space Time and Motion. New York: Braziller.

Koslow, Arnold. 1976.
"Ontological and ideological issues of the classical theory of space and time." In Peter K. Machamer and Robert G. Turnbull (eds.), Motion and Time, Space and Matter, pp. 224-263. Columbus, OH: Ohio State University Press.

Kozyren, N.A. 1971.
"On the possibility of experimental investigation of the properties of time." In Jiri Zeman (ed.), Time in Science and Philosophy, pp. 111-132. New York: Elsevier.

Kraft, R. Wayne. 1975.
"The mystery of time." In R. Wayne Kraft, Symbols, Systems, Science, and Survival, pp. 72-79. New York: Vintage.

Krips, H. 1971.
"The asymmetry of time." Australasian Journal of Philosophy, 49:204-210.

Kristofferson, Alfred B. 1967.
"Successiveness discrimination as a two-state quantal process." Science, 158:1337-1339.

Kroes, Peter. 1984.
"Objective versus mind-depedent theories of time flow." Synthese, 61:423-447.

Kronheimer, E.H., and R. Penrose. 1967.
"On the structure of causal spaces." Proceedings of the Cambridge Philosophical Society, 63:481-501.

Kummel, Friedrich. 1981.
"Time as succession and the problem of duration." In J.T. Fraser (ed.), The Voices of Time, 2nd edn., pp. 31-55. Amherst, MA: University of Massachusetts Press.

Kvale, S. 1974.
"The temporality of memory." Journal of Phenomenological Research, 5:7-31.

Lacaster, Lewis R. 1974.
"Discussion of time in Mahayana texts." Philosophy East and West, 24:209-214.

Lacey, H.M. 1968.
"The causal theory of time, a critique of Grunbaum's version." Philosophy of Science, 35:322-354.

Landsberg, P.T. 1972.
"Time in statistical physics and special relativity." In J.T. Fraser, F.C. Haber, and G.H. Muller (eds.), The Study of Time, pp. 59-109. New York: Springer-Verlag.

Landsberg, P.T. 1978.
"Thermodynamics, cosmology, and the physical constants." In J.T. Fraser, N. Lawrence, and D. Park (eds.), The Study of Time III, pp. 116-138. New York: Springer-Verlag.

Landsberg, P.T. 1982.
The Enigma of Time. Bristol, UK: A. Hilger.

Lawrence, Nathaniel. 1969.
"Time represented as space." Monist, 53:447-456.

Lawrence, Nathaniel. 1986.
"The origins of time." In J.T. Fraser, N. Lawrence, and F.C. Haber (eds.), Time, Science, and Society in China and the West: The Study of Time V, pp. 23-38. Amherst, MA: University of Masschusetts Press.

Layzer, David. 1975.
"The arrow of time." Scientific American, 233(6):56-69.

Lee, Harold N. 1963.
"Mead's doctrine of the past." Tulane Studies in Philosophy, 12:52-75.

Leibniz, Gottfried Wilhelm. 1975.
"Time as relational." In Charles M. Sherover, The Human Experience of Time, pp. 134-142. New York: New York University.

Levi, Albert William. 1959.
Philosophy and the Modern World. Bloomington, IN: Indiana University Press.

Lewis, David. 1979.
"Counterfactual dependence and time's arrow." Nous, 13:455-476.

Lewis, Wyndham. 1927.
Time and Western Man. Boston, MA: Beacon Hill.

Leyden, W. von. 1964.
"Time, number, and eternity in Plato and Aristotle." Philosophical Quarterly, 14:35-52.

Little, Jr., H. Ganse. 1974.
Decision and Responsibility: A Wrinkle in Time. Tallahassee, FL: American Academy of Religion.

Liu, Shu-Hsien. 1974.
"Time and temporality: the Chinese perspective." Philosophy East and West, 24:145-153.

Lloyd, G.E.R. 1976.
"Views on time in Greek thought." In Louis Gardet et al., Cultures and Time, pp. 117-148. Paris, France: Unesco Press.

Locke, John. 1975.
"The idea of duration." In Charles M. Sherover, The Human Experience of Time, pp. 122-133. New York: New York University Press.

Loeb, Louis E. 1974.
"Causal theories and causal overdetermination." Journal of Philosophy, 71:535.

Loizou, Andros. 1986.
The Reality of Time. Brookfield, VT: Gower.

Lombard, Lawrence Brian. 1982.
"Events and the essentiality of time." Canadian Journal of Philosophy, 12:1-17.

Lovejoy, Arthur O. 1922.
"Time, meaning and transcendence." Journal of Philosophy, 19:505-515.

Lucas, J.R. 1973.
A Treative on Time and Space. London, UK: Methuen.

Lucas, J.R. 1984.
Space, Time and Causality: An Essay in Natural Philosophy. Oxford, UK: Clarendon.

Lucas, J.R. 1987.
"The open future." In Raymond Flood and Michael Lockwood (eds.), The Nature of Time, pp. 125-134. New York: Basil Blackwell.

Lucas, J.R. 1989.
The Future: An Essay on God, Temporality and Truth. New York: Basil Blackwell.

Lynch, William F. 1985.
"Imagining past, present, future in one piece." Studies in Formative Spirituality, 6(1):65-72.

Mabbott, J.D. 1951.
"Our direct experience of time." Mind, 60:153-167.

Macbeath, Murray. 1983.
"Communication and time reversal." Synthese, 56:27-46.

Mackie, J.L. 1966.
"The direction of causation." Philosophical Review, 75:457.

Mackie, J.L. 1974.
The Cement of the Universe: A Study of Causation. Oxford, UK: Clarendon.

Mackinnon, Edward. 1962.
"Time and contemporary physics." International Philosophical Quarterly, 2:452.

MacKinnon, William J. 1973.
"An evolution inquiry into future human systems." Philosophy Forum, 13(1-2):95-109.

Madhavatirtha. 1951.
The Concept of Time in Indian Philosophy. Ahmedabad, India: Vedant Ashram.

Mahadevan, T.M.P. 1953.
Time and the Timeless. Madras, India: Upanishad Vihar.

Malament, David. 1977.
"Causal theories of time and the conventionality of simultaneity." Nous, 11(3):293-300.

Mandal, K.K. 1968.
A Comparative Study of the Concepts of Space and Time in Indian Thought. Varanasi, India: Chowkhamba Sanskrit Series Office.

Manders, K. 1982.
"On the space-time ontology of physical theories." Philosophy of Science, 49: 575-590.

Margenau, H. 1954.
"Can time flow backwards?" Philosophy of Science, 21:79-93.

Marrou, H.I. 1969.
Time and Timeliness. New York: Sheed and Ward.

Marsden, Dora. 1955.
The Philosophy of Time. Oxford, UK: Holywell Press.

Mathews, G. 1979.
"Time's arrow and the structure of spacetime." Philosophy of Science, 46:82-97.

Matsumoto, M. 1975.
"Time: being or consciousness alone? -- a realist view." In J.T. Fraser and N. Lawrence (eds.), The Study of Time II, pp. 206-215. New York: Springer-Verlag.

Mayo, Bernard. 1955.
"Professor Smart and temporal asymmetry." Australasian Journal of Philosophy, 33.

Mayo, Bernard. 1961.
"Objects, events and complementarity." Philosophical Review, 70:340-361.

Mayo, Bernard. 1962.
"The open future." Mind, 71:1-14.

Mays, W. 1972.
"Whitehead and the philosophy of time." In J.T. Fraser, F.C. Haber, and G.H. Muller (eds.), The Study of Time, pp. 354-369. New York: Springer-Verlag.

Mays, W. 1975.
"Temporality and time in Hegel and Marx." In J.T. Fraser and N. Lawrence (eds.), The Study of Time II, pp. 98-113. New York: Springer-Verlag.

Mbiti, J.S. 1970.
African Religions and Philosophy. Garden City, NY: Doubleday.

McCall, Storrs. 1966.
"Temporal flux." American Philosophical Quarterly, 3:270-281.

McCall, Storrs. 1971.
"Time and the physical modalities." In Eugene Freeman and Wifrid Sellars (eds.), Basic Issues in the Philosophy of Time, pp. 102-132. La Salle, IL: Open Court.

McCall, Storrs. 1976.
"Objective time flow." Philosophy of Science, 43:337-362.

McDermott, A. Charlene. 1976.
"The Sautrantika arguments against the Traikalyavada in the light of the contemporary tense revolution." Philosophy East and West, 24:193-200.

McDonald, D.B. 1927.
"Continuous re-creation and atomic time in Muslim scholastic theology." *Isis*, IX:326-44.

McGilvray, E.B. 1914.
"Time and the experience of time." *Philosophical Review*, 23:121-145.

McInerney, Peter K. 1983.
"My future, right or wrong." *Philosophical Studies*, 44:235-245.

McKeon, Richard. 1974.
"Time and temporality." *Philosophia East and West*, 24:123-128.

McLaughlin, Robert (ed.). 1982.
What? Where? When? Why?: Essays on Induction, Space and Time, Explanation. Dordrecht, Netherlands: Reidel.

McRae, Robert. 1979.
"Time and the monad." *Nature and System*, 1:103-109.

McTaggart, J.M.E. 1908.
"The unreality of time." *Mind*, 17(n.s.):457-474.

McTaggart, J.M.E. 1927.
The Nature of Existence. Cambridge, UK: Cambridge University Press.

McTaggart, J.M.E. 1978.
"Time." In Richard M. Gale (ed.), *The Philosophy of Time: A Collection of Essays*, pp. 86-97. Atlantic Highlands, NJ: Humanities.

McVittie, G.C. 1972.
"General relativity and time in the solar system." In J.T. Fraser, F.C. Haber, and G.H. Muller (eds.), *The Study of Time*, pp. 33-38. New York: Springer-Verlag.

Mead, George Herbert. 1980.
The Philosophy of the Present. Chicago, IL: University of Chicago Press.

Meerloo, Joost A.M. 1954.
The Two Faces of Man: Two Studies on the Sense of Time

and on Ambivalence. New York: International Universities Press.

Meerloo, Joost A.M. 1970.
Along the Fourth Dimension: Man's Sense of Time and History. New York: John Day Co.

Mehlberg, Henryk. 1961.
"Physical laws and time's arrow." In H. Feigl and G. Maxwell (eds.), Current Issues in the Philosophy of Science, pp. 105-138. New York: Holt, Rinehart, and Winston.

Mehlberg, Henryk. 1969.
"The problem of causality in an indeterministic science." International Journal of Theoretical Physics, 2:351-372.

Mehlberg, Henryk. 1971.
"Philosophical aspects of physical time." In Eugene Freeman and Wilfrid Sellars (eds.), Basic Issues In the Philosophy of Time, pp. 16-60. La Salle, IL: Open Court.

Meissner, W.W. 1967.
"The temporal dimension in the understanding of human experience." Journal of Existentialism, 7:129-160.

Mellor, D.H. 1981.
"McTaggart, fixity and coming trust." In Richard Healy (ed.), Reduction, Time and Reality, pp. 79-97. Cambridge, UK.

Mercier, Andre. 1978.
"Physical and metaphysical time." Epistemologia, 1:337-352.

Meredith, Patrick. 1972.
"The psychophysical structure of temporal information." In J.T. Fraser, F.C. Haber, and G.H. Muller (eds.), The Study of Time, pp. 259-273. New York: Springer-Verlag.

Merlan, P. 1947.
"Time consciousness in Husserl and Heidegger." Philosophy and Phenomenological Research, 8:23-54.

Merleau-Ponty, J. 1978.
"Ideas of beginnings and endings in cosmology." In J.T.

Fraser, N. Lawrence, and D. Park (eds.), The Study of Time III, pp. 333-350. New York: Springer-Verlag.

Michotte, A. 1963.
The Perception of Causality. London, UK: Methuen.

Miller, Izchak. 1982.
"Husserl's account of our temporal awareness." In Hubert Dreyfus (ed.), Husserl, Intentionality, and Cognitive Science. Cambridge, MA: MIT Press.

Miller, Izchak. 1984.
Husserl, Perception, and Temporal Awareness. Cambridge, MA: MIT Press.

Mink, Louis O. 1960.
"Time, McTaggart and Pickwickian language." Philosophical Quarterly, 10.

Minkowski, Eugene. 1933/1970.
Lived Time: Phenomenological and Psychopathological Studies. Evanston, IL: Northwestern University Press.

Minkowski, Eugene. 1975.
"The presence of the past." In Charles M. Sherover, The Human Experience of Time, pp. 504-518. New York: New York University Press.

Misner, C. 1969.
"Absolute zero of time." Physical Review, 186:1328-1333.

Mortensen, Chris. 1977.
"Critical notice." Australasian Journal of Philosophy, 55:149-157.

Moulyn, Adrian C. 1952.
"The functions of point and line in time measuring operations." Philosophy of Science, 19(2).

Moulyn, Adrian C. 1982.
The Meaning of Suffering: An Interpretation of Human Existence from the Viewpoint of Time. Westport, CT: Greenwood.

Mouton, David L. 1969.
"Thinking and time." Mind, 78:60-76.

Mundle, C.W.K. 1954.
"How specious is the specious present?" Mind, 63.

Mundle, C.W.K. 1959.
"Broad's view about time." In P.A. Schilpp (ed.), The Philosophy of C.D. Braod. La Salle, IL: Open Court.

Murray, Michael. 1981.
"Time in Hegel's Phenomemology of Spirit." The Review of Metaphysics, 34(4):682-705.

Murray, Rosalind. 1942.
Time and the Timeliness. London, UK: Centenary Press.

Murungi, John. 1980.
"Toward an African conception of time." International Philosophical Quarterly, 20:407-416.

Muses, C. 1967.
"Time, experience, and dimensionality: an introduction to higher kinds of number." Annals of the New York Academy of Sciences, 138:646-660.

Myers, Gerald. 1971.
"William James on time perception." Philosophy of Science, 38:353-360.

Nagel, E. 1965.
"Types of causal explanation in science." In D. Lerner (ed.), Cause and Effect, pp. 11-32. New York: Free Press.

Nakamura, Hajime. 1964.
Ways of Thinking of the Eastern Peoples: India, China, Tibet, Japan. Honolulu, HI: East-West Center Press.

Nakamura, Hajime. 1981.
"Time in Indian and Japanese thought." In J.T. Fraser (ed.), The Voices of Time, 2nd edn., pp. 77-91. Amherst, MA: University of Massachusetts Press.

Navon, David. 1978.
"On a conceptual hierarchy of time, space, and other dimensions." Cognition, 6:223-228.

Naylor, Margery Bedford. 1980.
"Fatalism and timeless truth." In Peter van Inwagen

(ed.), Time and Cause: Essays Presented to Richard Taylor, pp. 49-65. Dordrecht, Netherlands: Reidel.

Needham, Joseph. 1981.
"Time and knowledge in China and the West." In J.T. Fraser (ed.), The Voices of Time, 2nd edn., pp. 92-135. Amherst, MA: University of Massachusetts Press.

Nelson, Jack. 1972.
"Logically necessary and sufficient conditions for identity through time." American Philosophical Quarterly, 9:177-185.

Nerlich, Graham. 1976.
The Shape of Space. Cambridge, UK: Cambridge University Press.

Nerlich, Graham. 1979.
"Time and the direction of conditionship." Australian Journal of Philosophy, 57:14.

Newton-Smith, W.H. 1980.
The Structure of Time. London, UK: Routledge and Kegan Paul.

Newton-Smith, W.H. 1987.
"Space, time and space-time: a philosopher's view." In Raymond Flood and Michael Lockwood (eds.), The Nature of Time, pp. 22-35. New York: Basil Blackwell.

Nicholson, Graeme. 1986.
"Ecstatic temporality in Sein und Zeit." In Joseph J. Kockelmans (ed.), A Companion to Martin Heidigger's "Being and Time", pp. 208-226. Washington, DC: Center for Advanced Research in Phenomenolgy and University Press of America.

North, J.D. 1972.
"The time coordinate in Einstein's restricted theory of relativity." In J.T. Fraser, F.C. Haber, and G.H. Muller (eds.), The Study of Time, pp. 12-32. New York: Springer-Verlag.

Oaklander, L. Nathan. 1977.
"The 'timelessness' of time." Philosophy and Phenomenological Research, 38:228-233.

Oaklander, L. Nathan. 1983.
"McTaggart, Schlesinger, and the two-dimensional time hypothesis." Philosophical Quarterly, 33:391-397.

Oaklanader, L. Nathan. 1984.
Temporal Relations and Temporal Becoming: A Defense of a Russellian Theory of Time. Lanham, MD: University Press of America.

Oaklander, L. Nathan. 1985.
"A reply to Schlesinger's 'How to navigate the river of time'." Philosophical Quarterly, 35:93-94.

Odegard, Douglas. 1978.
"Phenomenal time." Ratio, 20:116-122.

Ogawa, H. 1978.
"The concept of time in Mithraic mysteries." In J.T. Fraser, N. Lawrence, and D. Park (eds.), The Study of Time III, pp. 658-679. New York: Springer-Verlag.

O'Hagan, Timothy. 1984.
"Must time have a stop?" Journal of the British Society for Phenomenology, 15:231-242.

Ono, Ken-Ichi. 1975.
"On the origin of indeterminacy." In J.T. Fraser and N. Lawrence (eds.), The Study of Time II, pp. 249-257. New York: Springer-Verlag.

Ono, Ken-Ichi. 1978.
"On two fundamental laws of nature and the role of time in time-space manifold." In J.T. Fraser, N. Lawrence, and D. Parks. (eds.), The Study of Time III, pp. 141-164. New York: Springer-Verlag.

Overseth, O.E. 1969.
"Experiments in time reversal." Scientific American, 221:89-101.

Owen, G.E.L. 1966.
"Plato and Parmenides on the timeless present." Monist, 50:317-340.

Owen, G.E.L. 1976.
"Aristotle on time." In Peter K. Machamer and Robert G.

Turnbull (eds.), Motion and Time, Space and Matter, pp. 3-27. Columbus, OH: Ohio State University Press.

Palter, Robert. 1973.
"Kant's formulation of the laws of motion." In Patrick Suppes (ed.), Space, Time and Geometry, pp. 94-14. Dordrecht, Netherlands: Reidel.

Pande, N.R.W. 1969.
Time, Space and Motion. Nagpur, India: Nagpur Vidyapeeth Mudranalaya.

Panikkar, Raimundo. 1972.
"The law of karma and the historical dimension of man." Philosophy East and West, 22(1):25-43.

Panikkar, Raimundo. 1974.
"Toward a typology of time and temporality in the ancient Indian tradition." Philosophy East and West, 24(2):161-164.

Panikkar, Raimundo. 1976.
"Time and history in the tradition of India: Kala and Karma." In Louis Gardet et al., Cultures and Time, pp. 63-88. Paris, France: Unesco Press.

Park, David. 1970.
"Are space and time necessary?" Scientia, 105:1-13.

Park, David. 1972.
"The myth of the passage of time." In J.T. Fraser, F.C. Haber, and G.H. Muller (eds.), The Study of Time, pp.110-121. New York: Springer-Verlag.

Park, David. 1975.
"Laws of physics and ideas of time." In J.T. Fraser and N. Lawrence (eds.), The Study of Time II, pp. 258-266. New York: Springer-Verlag.

Park, David. 1978.
"The past and the future." In J.T. Fraser, N. Lawrence, and D. Park (eds.), The Study of Time III, pp. 351-367. New York: Springer-Verlag.

Park, David. 1980.
The Image of Eternity: Roots of Time in the Physical World. Amherst, MA: University of Massachusetts Press.

Park, David. 1988.
The How and the Why: An Essay on the Origins and Development of Physical Theory. Princeton, NJ: Princeton University Press.

Park, David. 1989.
"Should physicists say that the past really happened?" In J.T. Fraser (ed.), Time and Mind: The Study of Time VI. Madison, CT: International Universities Press.

Parker, F.H. 1965.
"The temporal being of Western man." Review of Metaphysics, 18:629-646.

Patrides, C.A. (ed.). 1976.
Aspects of Time. Manchester, UK: Manchester University Press.

Pattaro, Germano. 1976.
"The Christian conception of time." In Louis Gardet et al., Cultures and Time, pp. 169-195. Paris, France: Unesco Press.

Pears, D.F. 1950-51.
"Time, truth, and inference." Proceedings of the Aristotelian Society, 51:1-24.

Penrose, O., and I.C. Percival. 1962.
"The direction of time." Proceedings of the Physical Society, 79:605.

Penrose, Roger. 1987.
"Big bangs, black holes and 'time's arrow'." In Raymond Flood and Michael Lockwood (eds.), The Nature of Time, pp. 36-62. New York: Basil Blackwell.

Pike, Nelson. 1970.
God and Timelessness. London, UK.

Piotrowski, S.L. 1979.
"Time and astrophysics: the time arrow and the expanding universe." In Frank Greenaway (ed.), Time and the Sciences, pp. 39-52. Paris, France: Unesco Press.

Plato. 1975.
"The creation of time." In Charles M. Sherover, The

Human Experience of Time, pp. 40-45. New York: New York University Press.

Plotinus. 1975.
"Time and engendered being." In Charles M. Sherover, The Human Experience of Time, pp. 63-80. New York: New York University Press.

Plotinus. 1978.
"Time and eternity." In Richard M. Gale (ed.), The Philosophy of Time: A Collection of Essays, pp. 24-37. Atlantic Highlands, NJ: Humanities.

Plumer, Gilbert. 1984.
"Time as success." International Studies in Philosophy, 16(1):35-55.

Plumer, Gilbert. 1984.
"Why time is extensive?" Mind, 93:265-270.

Plumer, Gilbert. 1985.
"The myth of the specious present." Mind, 94:19-35.

Poppel, Ernst. 1988.
Mindworks: Time and Conscious Experience. New York: Harcourt Brace Jovanovich.

Popper, Karl R. 1956.
"The arrow of time." Nature, 177:538.

Popper, Karl R. 1956.
"Irreversablity and mechanics." Nature, 178:382.

Popper, Karl R. 1957.
"Irreversible processes in physical theory." Nature, 179:1297.

Popper, Karl R. 1958.
"Irreversible processes in physical theory." Nature, 181:402-403.

Posner, Michael I. 1978.
Chronometric Explorations of Mind. Hillsdale, NJ: Erlbaum.

Post, E.J. 1979.
"The logic of time reversal." Foundations of Physics, 9:129-161.

Prigogine, Ilya. 1980.
From Being to Becoming: Time and Complexity in the Physical Sciences. San Francisco, CA: Freeman.

Prigogine, Ilya. 1984.
"The rediscovery of time." Zygon, 19:433-448.

Prior, A.N. 1953.
"Three-valued logics and future contingents." Philosophical Quarterly, 3.

Prior, A.N. 1957.
Time and Modality. Oxford, UK: Clarendon Press.

Prior, A.N. 1958.
"Time after time." Mind, 67:244-246.

Prior, A.N. 1959.
"Thank goodness that's over." Philosophy, 34.

Prior, A.N. 1967.
Past, Present, and Future. New York: Oxford University Press.

Prior, A.N. 1968.
"Now." Nous, 2(2):109-119.

Prior, A.N. 1968.
"'Now' corrected and condensed." Nous, 2(4):411-412.

Prior, A.N. 1968.
Papers on Time and Tense. Oxford, UK: Oxford University Press.

Prior, A.N. 1968.
"Time and change." Ratio, 10(2):173-177.

Prior, A.N. 1971.
"Recent advances in tense logic." In Eugene Freeman and Wilfrid Sellars (eds.), Basic Issues in the Philosophy of Time, pp. 1-15. La Salle, IL: Open Court.

Prior, A.N. 1972.
"The notion of the present." In J.T. Fraser, F.C. Haber, and G.H. Muller (eds.), The Study of Time, pp. 320-323. New York: Springer-Verlag.

Prokhovnik, Simon Jacques. 1978.
"Time as a universal property of nature." Epistemologia, 1:313-322.

Puligandla, R. 1974.
"Time and history in the Indian tradition." Philosophy East and West, 24:165-170.

Putnam, Hilary. 1962.
"It ain't necessarily so." Journal of Philosophy, 59.

Quinn, John M. 1960.
The Doctrine of Time in St. Thomas. Washington, DC: Catholic University Press.

Quinn, John M. 1979.
"The concept of time in Albert the Great." Southwestern Journal of Philosophy, 10(3):21-47.

Quinn, Philip L. 1979.
"Existence throughout an interval of time and existence at an instant of time." Ratio, 21(1):1-12.

Quinton, Anthony. 1962.
"Spaces and times." Philosophy, 37.

Reck, Andrew J. 1959.
"Bergson's theory of duration." Tulane Studies in Philosophy, 8:27-47.

Reichenbach, Hans. 1947.
Elements of Symbolic Logic. Berkeley, CA: University of California Press.

Reichenbach, Hans. 1956.
The Direction of Time. Berkeley, CA: University of California Press.

Reichenbach, Hans. 1958.
The Philosophy of Space and Time. New York: Dover.

Reid, Herbert. G. 1972.
"The politics of time: conflicting philosophical perspectives and trends." The Human Context, 4:456-483.

Reitmeister, Louis Aaron. 1962.
A Philosophy of Time. New York: Citadel Press.

Rennie, M.K. 1971.
"On postulates for temporal order." In Eugene Freeman and Wilfrid Sellars (eds.), Basic Issues in the Philosophy of Time, pp. 133-144. La Salle, IL: Open Court.

Rescher, Nicholas. 1978.
"Truth and necessity in temporal perspective." In Richard M. Gale (ed.), The Philosophy of Time: A Collection of Essays, pp. 183-220. Atlantic Highlands, NJ: Humanities.

Reyna, Ruth. 1971.
"Metaphysics of time in Indian philosophy and its relevance to particle science." In Jiri Zeman (ed.), Time in Science and Philosophy, pp. 227-239. New York: Elsevier.

Ricoeur, Paul. 1984.
Time and Narrative. Chicago, IL: University of Chicago Press.

Robb, A.A. 1921.
The Absolute Relations of Time and Space. Cambridge, UK: Cambridge University Press.

Robb, A.A. 1936.
A Theory of Time and Space, 2nd edn. Cambridge, UK: Cambridge University Press.

Robb, A.A. 1936.
The Geometry of Time and Space. Cambridge, UK: Cambridge University Press.

Roberts, Joy H. 1985.
"On Russell's rejection of akoluthic sensations." Philosophy and Phenomenological Research, 45(4):595-600.

Rosen, D. 1978.
"In defense of a probabilistic theory of causality." Philosophy of Science, 45:604-613.

Rosen, D. 1982.
"A critique of deterministic causality." Philosophical Forum, 14(2):101-130.

Rouner, Leroy S. 1974.
"Creationism and emanationism: a problem in Radhakrishnan's philosophy." Philosophy East and West, 24:227-238.

Roxburgh, Ian W. 1977.
"The conventionality of uniform time." British Journal for the Philosophy of Science, 28:172-177.

Royce, Josiah. 1910.
"The reality of the temporal." International Journal of Ethics, 20:257-271.

Russell, Bertrand. 1915.
"On the experience of time." Monist, 25:212-233.

Russell, Bertrand. 1936.
"On order in time." Proceedings of the Cambridge Philosophical Society, 32:216-228.

Russell, Henry Norris. 1949.
"The time scale of the universe." In Time and its Mysteries, Series III, pp. 3-30. New York: New York University Press.

Russell, J.L. 1981.
"Time in Christian thought." In J.T. Fraser (ed.), The Voices of Time, 2nd edn., pp. 59-76. Amherst, MA: University of Massachusetts Press.

Ryan, John K. (ed.).
The Concept of Time in St. Augustine and Other Studies. Baltimore, MD: Catholic University of America Press.

Sachs, Robert G. 1963.
"Can the direction of flow of time be determined?" Science, 140:1284-1290.

Sachs, Robert G. 1987.
The Physics of Time Reversal. Chicago, IL: University of Chicago Press.

Salmon, Wesley C. 1976.
"Clocks and simultaneity in special relativity or, Which twin has the Timex." In Peter K. Machamer and Robert G. Turnbull (eds.), Motion and Time, Space and Matter, pp. 508-545. Columbus OH: Ohio State University Press.

Salmon, Wesley C. 1980.
"Probabilistic causality." Pacific Philosophical Quarterly, 61:50-74.

Salmon, Wesley C. 1980.
Space, Time, and Motion: A Philosophical Introduction, 2nd edn. Minneapolis, MN: University of Minnesota Press.

Sanford, David H. 1976.
"The direction of causation and the direction of conditionship." Journal of Philosophy, 73:193-207.

Sanford, David H. 1984.
"The direction of causation and the direction of time." In Peter A. French, Theodore W. Uehling, Jr., and Howard K. Wettstein (eds.). Causation and Causal Theories, pp. 53-75. Minneapolis, MN: University of Minnesota Press.

Santamaria, Ulysses, and Anne M. Bailey. 1984.
"A note on Braudel's structure as duration." History and Theory, 23(1):78-83.

Santayana, George. 1975.
"Sentimental time." In Charles M. Sherover, The Human Experience of Time, pp. 407-418. New York: New York University Press.

Savile, Anthony. 1982.
The Test of Time: An Essay in Philosophical Aesthetics. New York: Oxford University Press.

Schaffner, Kenneth F. 1976.
"Space and time in Lorentz, Poincare, and Einstein: divergent approaches to the discovery and development of the special theory of relativity." In Peter K. Machamer and Robert G. Turnabull (eds.), Motion and Time, Space

and Matter, pp. 465-507. Columbia, OH: Ohio State University Press.

Schaltenbrand, Georges. 1967.
"Consciousness and time." Annals of the New York Academy of Sciences, 138:632-645.

Schlegel, Richard. 1981.
"Time and thermodynamics." In J.T. Fraser (ed.), The Voices of Time, 2nd edn., pp. 500-523. Amherst, MA: University of Massachusetts Press.

Schlesinger, George N. 1980.
Aspects of Time. Indianapolis, IN: Hackett.

Schlesinger, George N. 1982.
"How time flies." Mind, 91:501-523.

Schlesinger, George N. 1985.
"How to navigate the river of time." Philosophical Quarterly, 35:91-92.

Schrodinger, E. 1950.
"Irreversibility." Proceedings of the Royal Irish Academy, 52.

Schuster, Melvin M. 1961.
"An analysis of relational time." Review of Metaphysics, 15:209-224.

Schuster, Melvin M. 1968.
"On the denial of past and future existence." Review of Metaphysics, 21:447-467.

Schuster, Melvin M. 1986.
"Is the flow of time subjective?" Review of Metaphysics, 39:695-714.

Schutz, Alfred. 1967.
The Phenomenology of the Social World. Evanston, IL: Northwestern University Press.

Schwartz, Adria E. 1977.
"Being-in-time: a phenomenological exploration of the existential past, present and future." Review of Existential Psychology and Psychiatry, 15(2-3):150-160.

Sciama, Dennis. 1987.
"Time 'paradoxes' in relativity." In Raymond Flood and Michael Lockwood (eds.), The Nature of Time, pp. 6-21. New York: Basil Blackwell.

Sebba, G. 1972.
"Time and the modern self: Descartes, Rousseau, Beckett." In J.T. Fraser, F.C. Haber, and G.H. Muller (eds.), The Study of Time, pp. 452-469. New York: Springer-Verlag.

Sellars, Wilfrid. 1962.
"Philosophy and the scientific image of man." In Robert G. Colodny (ed.), Frontiers of Science and Philosophy, pp. 35-78. Pittsburgh, PA: University Pittsburgh Press.

Sellars, Wilfrid. 1962.
"Time and the world order." In H. Feigl and G. Maxwell, Scientific Explanation, Space and Time, pp. 527-616. Minneapolis, MN: University of Minnesota Press.

Sellars, Wilfrid. 1962.
"Time and the world order." In H. Feigl and G. Maxwell (eds.), Minnesota Studies in the Philosophy of Science, Vol. 3. Minneapolis, MN: University of Minnesota Press.

Sen, Sanat Kumar. 1968.
"Time in Sankhya-Yoga." International Philosophical Quarterly, 8:406-426.

Shallis, Michael. 1983.
On Time: An Investigation Into Scientific Knowledge and Human Experience. New York: Schocken.

Shallis, Michael. 1987.
"Time and cosmology." In Raymond Flood and Michael Lockwood (eds.), The Nature of Time, pp. 63-79. New York: Basil Blackwell.

Shalom, Albert. 1989.
"Time and mind in the constitution of the universe: a critique of the participatory anthropic principle." In J.T. Fraser (ed.), Time and Mind: The Study of Time VI. Madison, CT: International Universities Press.

Shapiro, H. 1957.
Motion, Time and Space According to William of Ockham. St. Bonaventure, New York: Fransican Institute.

Sharvy, Richard. 1971.
"Things." In Eugene Freeman and Wilfrid Sellars (eds.), Basic Issues in the Philosophy of Time, pp. 164-180. La Salle, IL: Open Court.

Sherover, Charles M. 1971.
Heidegger, Kant and Time. Bloomington, IN: Indiana University Press.

Sherover, Charles M. 1975.
The Human Experience of Time: The Development of Its Philosophic Meaning. New York: New York University Press.

Sherover, Charles M. 1975.
"Time and ethics: how is morality possible?" In J.T. Fraser and N. Lawrence (eds.), The Study of Time II, pp. 216-230. New York: Springer-Verlag.

Sherover, Charles M. 1986.
"Are we in time?" International Philosophical Quarterly, 26:33-46.

Sherover, Charles M. 1989.
"Res cogitans: the time of mind." In J.T. Fraser (ed.), Time and Mind: The Study of Time VI. Madison, CT: International Universities Press.

Shoemaker, Sydney S. 1969.
"Time without change." Journal of Philosophy, 66:363-381.

Shoemaker, Sydney S. 1970.
"Persons and their pasts." American Philosophical Quarterly, 7:269-285.

Shorter, J.M. 1984.
"The reality of time." Philosophia, 14:321-339.

Shwayder, D.S. 1957.
"The temporal order." Philosophical Quarterly, 7:32-43.

Siffre, M. 1964.
Beyond Time. New York: McGraw-Hill.

Sipfle, David A. 1971.
"On the intelligibility of the epochal theory of time."

In Eugene Freeman and Wilfrid Sellars (eds.), Basic Issues in the Philosophy of Time, pp. 181-194. La Salle, IL: Open Court.

Sivin, Nathan. 1976.
"Chinese alchemy and the manipulation of time." Isis, 67(239):513-526.

Sivin, Nathan. 1986.
"On the limits of empirical knowledge in the traditional Chinese sciences." In J.T. Fraser, N. Lawrence, and F.C. Haber (eds.), Time, Science, and Society in China and the West: The Study of Time V, pp. 151-169. Amherst, MA: University of Massachusetts Press.

Sklar, Lawrence. 1977.
Space, Time and Spacetime. Berkeley, CA: University of California Press.

Sklar, Lawrence. 1977.
"Facts, conventions, and assumptions in the theory of spacetime." In John S. Earman, Clark N. Glymour, and John J. Stachel (eds.), Foundations of Space-Time Theories, pp. 206-274. Minneapolis, MN: University of Minnesota Press.

Sklar, Lawrence. 1977.
"What might be right about the causal theory of time." Synthese, 35:155-171.

Sklar, Lawrence. 1981.
"Up and down, left and right, past and future." Nous, 15:111-129.

Sklar, Lawrence. 1981.
"Time, reality, and relativity." In R. Healy (ed.), Reduction, Time, and Reality, pp. 129-142. Cambridge, UK: Cambridge University Press.

Sklar, Lawrence. 1983.
"Prospects for a causal theory of space-time." In R. Swinburne (ed.), Space, Time, and Causality, pp. 45-62. Dordrecht, Netherlands: Reidel.

Sklar, Lawrence. 1985.
Philosophy and Spacetime Physics. Berkeley, CA: University of California Press.

Slaate, Howard A. 1989.
Time, Existence and Destiny. New York: Peter Lang Publishing.

Smart, J.J.C. 1949.
"The river of time." Mind, 58:483-494.

Smart, J.J C. 1952-53.
"The temporal asymmetry of the world." Analysis, 14.

Smart, J.J.C. 1955.
"Mr. Mayo on temporal asymmetry." Australasian Journal of Philosophy, 33.

Smart, J.J.C. 1955.
"Spatialising time." Mind, 64:239-241.

Smart, J.J.C. 1963.
Philosophy and Scientific Realism. London: Rontelege and Kegan Paul.

Smart. J.J.C. 1963.
"Is time travel possible?" Journal of Philosophy, 60.

Smart, J.C.C. (ed.). 1964.
Problems of Space and Time. New York: Macmillan.

Smart, J.J.C. 1968.
Between Science and Philosophy. New York: Random House.

Smart, J.J.C. 1971.
"Causal theories of time." In Eugene Freeman and Wilfrid Sellars (eds.), Basic Issues in the Philosophy of Time, pp. 61-71. La Salle, IL: Open Court.

Smart, J.J.C. 1980.
"Time and becoming." In Peter van Inwagen (ed.), Time and Cause: Essays Presented to Richard Taylor, pp. 3-15. Dordrecht, Netherlands: Reidel.

Smart, J.J.C. 1981.
"The reality of the future." Philosophia, 10:141-150.

Smith, John E. 1969.
"Time, times, and the 'right time': chronos and kairos." Monist, 53:1-13.

Smith, John E. 1986.
"Time and qualitative time." Review of Metaphysics, 40:3-16.

Smith, J. Maynard. 1972.
"Time in the evolutionary process." In J.T. Fraser, F.C. Haber, and G.H. Muller (eds.), The Study of Time, pp. 200-206. New York: Springer-Verlag.

Sokolowski, Robert. 1982.
"Timing." Review of Metaphysics, 35:687-714.

Sorabji, Richard. 1983.
Time, Creation and the Continuum: Theories in Antiquity and the Early Middle Ages. Ithaca, NY: Cornell University Press.

Sosa, Ernest (ed.). 1975.
Causation and Conditionals. Oxford, UK: Oxford University Press.

Spellman, Lynne. 1981.
"Forthcoming sea fights: what theory of time?" New Scholasticism, LV(1):52-68.

Spenner, M. 1982.
"Social sciences and the consciousness of the future." Theory and Society, 11(5):683-712.

Stambaugh, Joan. 1974.
"Time, finitude, and finality." Philosophy East and West, 24:129-135.

St. Augustine. 1978.
"Some questions about time." In Richard M. Gale (ed.), The Philosophy of Time: A Collection of Essays, pp. 38-54. Atlantic Highlands, NJ: Humanities.

Stephenson, William. 1988.
"William James, Niels Bohr, and complementarity: IV. The significance of time." Psychological Record, 38(1):19-35.

Stevens, Jr., Herbert H. 1980.
"Reality, finitism, and time." Speculations in Science and Technology, 3:587-593.

Straus, Erwin W. 1967.
"An existence approach to time." Annals of the New York Academy of Sciences, 138:759-766.

Stump, Eleanore, and Norman Kretzmann. 1981.
"Eternity." Journal of Philosophy, 78:429-458.

Sugarman, R. 1980.
Rancor Against Time: The Phenomenology of 'Ressentiment'. Hamburg, FRG: Felix Meiner Verlag.

Suppes, Patrick. 1970.
A Probabilistic Theory of Causality. Amsterdam, Netherlands: North Holland.

Suppes, Patrick, (ed.). 1973.
Space, Time and Geometry. Dordrecht, Netherlands: Reidel.

Suppes, Patrick. 1973.
"Some open problems in the philosophy of space and time." In Patrick Suppes (ed.), Space, Time and Geometry, pp. 383-401. Dordrecht, Netherlands: Reidel.

Suppes, Patrick. 1984.
"Conflicting intuitions about causality." In Peter A. French, Theodore E. Uehling, Jr., and Howard K. Wettstein (eds.), Causation and Causal Laws, pp. 151-168. Minneapolis, MN: University of Minnesota Press.

Surber, Jere Paul. 1979.
"Heidegger's critique of Hegel's notion of time." Philosophy and Phenomenological Research, 39:358-377.

Swinburne, Richard G. 1968.
Space and Time. London, UK: Macmillan.

Swinburne, Richard G. 1980.
"Conventionalism about space and time." British Journal for the Philosophy of Science, 31(3):255-272.

Swinburne, Richard G. (ed.). 1983.
Space, Time and Causality. Dordrecht, Netherlands: Reidel.

Swoyer, Chris. 1984.
"Causation and identity." In Peter A. French, Theodore

E. Uehling, Jr., and Howard K. Wettstein (eds.), Causation and Causal Theories, pp. 593-622. Minneapolis, MN: University of Minnesota Press.

Szamosi, Gesa. 1986. The Twin Dimensions: Inventing Time and Space. New York: McGraw-Hill.

Szumilewicz, Irena. 1971. "The direction of time and entropy." In Jiri Zeman (ed.), Time in Science and Philosophy, pp. 181-192. New York: Elsevier.

Taschdjian, Edgar. 1977. "Time horizon: the moving boundary." Behavioral Science. 22:41-48.

Tatarkiewicz, W. 1966. "Happiness and time." Philosophy and Phenomenological Research, 27(1):1-10.

Taylor, J.G. 1972. "Time in particle physics." In J.T. Fraser, F.C. Haber, and G.H. muller (eds.) The Study of Time, pp. 53-58. New York: Springer-Verlag.

Taylor, Richard. 1955. "Spatial and temporal analogies and the concept of identity." Journal of Philosophy, 52:599-612.

Taylor, Richard. 1957. "The problem of future contingents." Philosophical Review, 66:18.

Taylor, Richard. 1959. "Moving about in time." Philosophical Quarterly, 9:289-301.

Taylor, Richard. 1962. "Fatalism and ability." Analysis, 23:25-27.

Taylor, Richard. 1962. "Fatalism." Philosophical Review, 71:56-66.

Taylor, Richard. 1963. "A note on fatalism." Philosophical Review, 72:497-499.

Taylor, Richard. 1966.
"Prevention, postvention and the will." In Keith Lehrer (ed.), Freedom and Determinism, pp. 65-85. New York: Random House.

Teske, Roland J. 1983.
"The world-soul and time in St. Augustine." Augustine Studies, 14:75-92.

Thalberg, Irving. 1971.
"Free will and Chisholm's varieties of causation." Idealistic Studies, 1:149-159.

Thalberg, Irving. 1973.
"Constituents and causes of emotion and action." Philosophical Quarterly, 13:1-14.

Thalberg, Irving. 1980.
"Fatalism toward past and future." In Peter van Inwagen (ed.), Time and Cause: Essays Presented to Richard Taylor, pp. 27-47. Dordrecht, Netherlands: Reidel.

Thomason, R. 1984.
"Combinations of tense and modality." In D. Gabbay and F. Guenthner (eds.), Handbook of Philosophical Logic, Vol. II, pp. 135-165. Dordrecht, Netherlands: Reidel.

Thomason, Richmond H. 1970.
"Indeterminist time and truth-value gaps." Theoria, 36(3):264-281.

Toda, Masanao. 1978.
"The boundaries of the notion of time." In J.T. Fraser, N. Lawrence, and D. Park (eds.), The Study of Time III, pp. 370-388. New York: Springer-Verlag.

Tooley, Michael. 1984.
"Laws and causal relations." In Peter A. French, Theodore E. Uehling, Jr., and Howard K. Wettstein (eds.), Causation and Causal Theories, pp. 93-112. Minneapolis, MN: University of Minnesota Press.

Trefil, James F. 1985.
Space, Time, Infinity. Washington, DC: Smithsonian Books.

Trivers, Howard. 1985.
The Rhythm of Being: A Study of Temporality. New York: Philosophical Library.

Tu, Wei-Ming. 1974.
"An introductory note on time and temporality." Philosophy East and West, 24:119-122.

Tymieniecka, Anna-Teresa. 1962.
Phenomenology and Science in Contemporary Human Thought. New York: Farrar, Straus and Giroux.

Van Der Leeuw, G. 1957.
"Primordial time and final time." In J. Campbell (ed.), Man and Time. New York: Pantheon Books.

Van Fraassen, Bas C. 1962.
"Capek on eternal recurrence." Journal of Philosophy, 59:371-375.

Van Fraassen, Bas C. 1973.
"Earman on the causal theory of time." In Patrick Suppes (ed.), Space, Time and Geometry, pp. 85-93. Dordrecht, Netherlands: Reidel.

Van Fraassen, Bas C. 1978.
"Time: physical and experienced." Epistemologia, 1:323-336.

Van Fraassen, Bas, C. 1982.
"Rational belief and the common cause principle." In Robert McLaughlin (ed.), What? Where? When? Why?: Essays on Induction, Space and Time, Explanation, pp. 193-209. Dordrecht, Netherlands: Reidel.

Van Fraassen, Bas C. 1985.
An Introduction to the Philosophy of Time and Space. New York: Columbia University Press.

Van Inwagen, Peter (ed.). 1980.
Time and Cause: Essays Presented to Richard Taylor. Dordrecht, Netherlands: D. Reidel.

Vendler, Zeno. 1957.
"Verbs and times." Philosophical Review, 66.

Voegelin, Eric. 1972.
"On Hegel: a study in sorcery." In J.T. Fraser, H.C. Haber, and G.H. Muller (eds.), The Study of Time, pp. 418-451. New York: Springer-Verlag.

Von Leyden, W. 1964.
"Time, number and eternity in Plato and Aristotle." Philosophical Quarterly, 14:35-52.

Von Wright, George Henrik. 1968.
Time, Change and Contradiction. Cambridge, UK: Cambridge University Press.

Von Wright, George Henrik. 1974.
Causality and Determinism. New York: Columbia University Press.

Wagn, Klaus. 1976.
What Time Does. Munich, FRG: Caann Verlag.

Waismann, Friedrich. 1950.
"Analytic-synthetic." Analysis, 11(2):25-38.

Wald, Robert M. 1977.
Space, Time and Gravity. Chicago: University of Chicago Press.

Wallis, Robert. 1968.
Time: Fourth Dimension of the Mind. New York: Harcourt, Brace and World.

Waltstad, A. 1980.
"Time's arrow in an oscillating universe." Foundations of Physics, 10:743-750.

Watanabe, M.S. 1975.
"Causality and Time." In J.T. Fraser, and N. Lawrence (Eds.), The Study of Time II, pp. 267-282. New York: Springer-Verlag.

Watanabe, Satosi. 1972.
"Creative time." In J.T. Fraser, F.C. Haber, G.H. Muller (eds.), The Study of Time, pp. 159-189. New York: Springer-Verlag.

Watanabe, Satosi. 1981.
"Time and the probabilitistic view of the world." In

J.T. Fraser. (ed.), The Voices of Time, 2nd edn, pp. 527-563. Amherst, MA: University of Massachusetts Press.

Waterlow, Sarah. 1974.
"Backwards causation and continuing." Mind, 83:372-287.

Webb, C.W. 1960.
"Could time flow, if so how fast?" Journal of Philosophy, 57.

Weinberg, Steven. 1976.
The First Three Minutes: A Modern View of the Origin of the Universe. New York: Basic.

Weingard, Robert. 1973.
"On traveling backward in time." In Patrick Suppes (ed.), Space, Time and Geometry, pp. 115-130. Dordrecht, Netherlands: Reidel.

Weingard, Robert. 1977.
"Space-time and the direction of time." Nous, 11:119-132.

Wesley, J.P. 1980.
"Is 'time dilation' only approximately valid?" Speculations in Science and Technology, 3:409-417.

White, M. 1984.
"The necessity of the past and modal-tense logic incompleteness." Notre Dame Journal of Formal Logic, 25:69-71.

Whitehead, Alfred North. 1965.
"Time." In D. Browning (ed.), Philosophers of Process. New York: Random House.

Whitehead, Alfred North. 1975.
"Two kinds of time relatedness." In Charles M. Sherover, The Human Experience of Time, pp. 330-346. New York: New York University Press.

Whiteman, Michael. 1967.
Philosophy of Space and Time. London, UK: Allen and Unwin.

Whitrow, Gerald J. 1967.
"Reflections on the natural philosophy of time." Annals of the New York Academy of Sciences, 138:422-432.

Whitrow, Gerald J. 1972.
"Reflections on the history of the concept of time." In J.T. Fraser, F.C. Haber, and G.H. Muller (eds.), The Study of Time. pp. 1-11. New York: Springer-Verlag.

Whitrow, Gerald J. 1972.
What is Time? London, UK: Thames and Hudson.

Whitrow, Gerald J. 1973.
The Nature of Time. New York: Holt, Rinehart and Winston.

Whitrow, Gerald J. 1978.
"On the impossibility of an infinite past." British Journal for the Philosophy of Science, 29:39-45.

Whitrow, Gerald J. 1979.
"Introduction." In Frank Greenaway (ed.), Time and the Science, pp. 1-20. Paris, France: Unesco Press.

Whitrow, Gerald J. 1979.
"Mathematical time and its role in the development of the scientific world-view." In Frank Greenaway (ed.), Time and the Sciences, pp. 21-37. Paris, France: Unesco Press.

Whitrow, Gerald J. 1980.
The Natural Philosophy of Time, 2nd edn. Oxford, UK: Clarendon.

Whitrow, Gerald J. 1981.
"Time and the universe." In J.T. Fraser (ed.), The Voices of Time, 2nd edn., pp. 564-581. Amherst, MA: University of Massachusetts Press.

Whitrow, Gerald J. 1989.
"Man and time: some historical and critical reflections." In J.T. Fraser (ed.), Time and Mind: The Study of Time VI. Madison, CT: International Universities Press.

Whittemore, Robert. 1955.
"Time and Whitehead's God." Tulane Studies in Philosophy, 4:83-92.

Whyte, Lancelot Law. 1972.
"Pierre Curie's principle of one way process." In J.T. Fraser, F.C. Haber, and G.H. Muller (eds.), The Study of Time, pp. 140-147. New York: Springer-Verlag.

Wiener, Norbert. 1961.
"Newtonian time and Bergsonian time." In Norbert Wiener, Cybernetics, 2nd edn. New York: Wiley.

Wiener, T. 1958.
"Time and the science of organization." Scientia, 93:199-205.

Wiggins, David. 1971.
Identity and Spatio-Temporal Continuity. Oxford, UK.

Wilcox, Donald J. 1987.
The Measure of Times Past: Pre-Newtonian Chronologies and the Rhetoric of Relative Time. Chicago, IL: University of Chicago Press.

Will, Frederick L. 1965.
"Will the future be like the past?" In Anthony Flew, Logic and Language, pp. 248-266. Garden City, NY: Anchor Press.

Williams, Donald C. 1951.
"The sea fight tomorrow." In P. Henle et al. (eds.), Structure, Method and Meaning. New York: Liberal Arts Press.

Williams, Donald C. 1951.
"The myth of passage." Journal of Philosophy, 48:457-472.

Wilson, N.L. 1955.
"Space, time and individuals." Journal of Philosophy, 52.

Winnie, John (ed.). 1977.
"Symposium on space and time." Nous, 11(3).

Winnie, John. 1977.
"The causal theory of space-time." In John S. Earman, Clark N. Glymour, and John T. Stachel (eds.), Foundations of Space-Time Theories, pp. 134-205. Minneapolis, MN: University of Minnesota Press.

Wold, Herman O.A. 1967.
"Time as the realm of forecasting." Annals of the New York Academy of Sciences, 138:525-560.

Wood, David. 1982.
"Time and the sign." Journal of the British Society for Phenomenology, 13:143-153.

Wood, David. 1988.
The Deconstruction of Time. Atlantic Highlands, NJ: Humanities.

Wood, David, and Robert Bernasconi (eds.). 1982.
Time and Metaphysics: A Collection of Original Papers. Parousia Press.

Yaker, Henri M. 1972.
"Time in the Biblical and Greek worlds." In Henri Yaker, Humphry Osmond, and Frances Cheek (eds.), The Future of Time: Man's Temporal Environment, pp. 15-35. Garden City, NY: Doubleday.

Yamamoto, M. 1975.
"What time is not." In J.T. Fraser and N. Lawrence (eds.), The Study of Time II, pp. 231-238. New York: Springer-Verlag.

Yinzhi, Zhang. 1986.
"Mohist views of time and space: a brief analysis." In J.T. Fraser, N. Lawrence, and F.C. Haber (eds.), Time, Science, and Society in China and the West: The Study of Time V, pp. 206-210, Amherst, MA: University of Massachusetts Press.

Yourgrau, Palle. 1985.
"On the logic of indeterminist time." Journal of Philosophy, 82:548-559.

Zeman, Jiri. 1971.
"The concept of gradient channel in ontology and epistemology and the problem of time." In Jiri Zeman (ed.), Time in Science and Philosophy, pp. 193-207. New York: Elsevier.

Zimmerman, E.J. 1981.
"Time and quantum theory." In J.T. Fraser (ed.), The

Voices of Time, 2nd edn., pp. 479-499. Amherst, MA: University of Massachusetts Press.

Zimmerman, Michael E. 1977.
"Heidegger and Nietzsche on authentic time." Cultural Hermeneutics, 4(3):239-264.

Zwart, P.J. 1973.
"The flow of time." In Patrick Suppes (ed.), Space, Time and Geometry, pp. 131-156. Dordrecht, Netherlands: Reidel.

Zwart, P.J. 1976.
About Time: A Philosophical Inquiry Into the Origin and Nature of Time. New York: American Elsevier.

15
Physiology

Andrew, R.J., and A. Brennan. 1985.
"Sharply timed and lateralised events at time of establishment of long term memory." Physiology and Behavior, 34(4):547-556.

Bell, C.R., and K.A. Provins. 1963.
"Relations between physiological responses to environmental heat and time judgments." Journal of Experimental Psychology, 66:572-579.

Bunning, Erwin. 1973.
The Physiological Clock, rev. 3rd edn. New York: Springer-Verlag.

Cahoon, Richard L. 1969.
"Physiological arousal and time estimation." Perceptional and Motor Skills, 28:259-268.

Caldwell, John A. and Gary E. Jones. 1985.
"The effects of exposure to red and blue light on physiological indices and time estimation." Perception, 14(1):19-29.

Carlson, V.R., D.R. Goodenough, and I. Fineberg. 1968.
"Time perception and sleep." Psychophysiology, 5:242.

Contreras, Carlos M., Lilian Mayagoitia, and Graciela Mexicano. 1985.
"Interemispheric changes in alpha rhythm related to time perception." Physiology and Behavior, 34(4):525-529.

Doehring, D.G., J.E. Helmer, and Elizabeth A. Fuller. 1966.
"Physiological responses associated with time estimation in a human operant situation." Psychological Record, 14:355-362.

Fernandez-Guardiola, Augusto, L. Jose Juardo, and Efrian Aguilar-Jimenez. 1984.
"Evaluation of the attention and sleepiness states by means of a psychophysiological test of reaction time and time estimate in man: effects of psychotropic drugs." Current Therapeutic Research, 35(6):1000-1009.

Halberg, F. 1969.
"Chronobiology." Annual Review of Physiology, 31:675-725.

Hartmann, E. 1968.
"The day residue: time distribution of waking events." Psychophysiology, 5(2):222.

Hoagland, Hudson. 1933.
"The physiological control of the judgment of duration: evidence for chemical clock." Journal of General Psychology, 9:267-287.

Kleber, R.J., W.T. Lhamon, and S. Goldstone. 1963.
"Hyperthermia, huperthyroidism, and time judgment." Journal of Comparative Physiological Psychology, 56:362-365.

Klein, K.E., H. Bruner, et al. 1972.
"Psychological and physiological changes caused by desynchronization following transzonal air travel." In W.P. Colquhoun (ed.), Aspects of Human Efficiency, pp. 295-305. London, UK: English Universities Press.

Lavie, P., and W.B. Webb. 1975.
"Time estimation in a long-term time-free environment." American Journal of Physiology, 88:177-186.

Luce, Gay Gaer. 1971.
Body Time: Physiological Rhythms and Social Stress. New York: Pantheon.

Mills, J.N. 1964.
"Circadian rhythms during and after three months in solitude underground." Journal of Physiology, 174:217-231.

Milner, B. 1966.
"Amnesia following operation on the temporal lobe." In C.M. Whitty and O.L. Zangwill (eds.), Amnesia, pp. 112-115. London, UK: Butterworths.

Milner, B. 1972.
"Disorders of learning and memory after temporal lobe lesions in man." Clinical Neurosurgery, 19:421-446.

Minors, D.S., and J.M. Waterhouse. 1981.
Circadian Rhythms and the Human. Boston, MA: J. Wright.

Moore-Ede, M.C., F.M. Sulzman, and C.A. Fuller. 1982.
The Clocks That Time Us: Physiology of the Circadian Timing Sytem. Cambridge, MA: Harvard University Press.

Oswald, Ian. 1967.
"Slow neurophysiological swings." Annals of the New York Academy of Sciences, 138:616-622.

Poppel, Ernst. 1976.
"Time perception." In H.L. Tenber, R. Held, and H. Leibowitz (eds.), Handbook of Sensory Physiology, Vol. 8, pp. 713-729. New York: Springer-Verlag.

Poppel, Ernst. 1987.
"Time perception." In George Adelman (ed.), Encyclopedia of Neuroscience, 2nd edn., pp. 1215-1216. Boston. MA: Birkhauser.

Raevskaya, O.S., and T.D. Dzhebrailova. 1987.
"Dynamics of physiological parameters during perception of minute time intervals." Human Physiology, 13(2):90-96.

Richter, C.P. 1957.
"Hormones and rhythms in man and animals." Hormone Research, 13:105-159.

Rusak, B., and I. Zucker. 1979.
"Neural regulation of circadian rhythms." Physiological Review, 59:449-526.

Stephan, F.K. 1984.
"Phase shifts of circadian rhythms in activity entrained to food access." Physiology and Behaviour, 32:663-672.

Stolerman, I.P. 1985.
"Cue location does not account for all selective associations in food aversion learning." Quarterly Journal of Experimental Psychology: Comparative and Physiological Psychology, 37(2B):181-185.

Strughold, H. 1965.
"The physiological clock in aeronautics and astronautics." Annals of the New York Academy of Sciences, 134:413-422.

Sulzman, F.M., C.A. Fuller, and M.C. Moore-Ede. 1978.
"Comparison of synchronization of primate circadian rhythm by light and food." American Journal of Physiology, 234:R130-R135.

Wirz-Justice, A., G.A. Groos, and T.A. Wehr. 1982.
"The neuropharmacology of circadian timekeeping in mammals." In J. Aschoff, S. Daan, and G.A. Gross (eds.), Vertebrate Circadian Systems: Structure and Physiology, pp. 183-193. New York: Springer-Verlag.

Woodruff, D. 1975.
"Relationships among EEG alpha frequency, reaction time, and age: a biofeedback study." Psychophysiology, 12:673-681.

16
Political Science

Back, Kurt. W., and Kenneth J. Gergen. 1963.
"Apocalyptic and serial time orientations and the structure of opinions." Public Opinion Quarterly, 27: 427-442.

Gunnel, John, G. 1968.
Political Philosophy and Time. Middletown, CT: Wesleyan University Press.

Gunnel, John, G. 1975.
"The history of political philosophy and the myth of the tradition." In J.F. Fraser and N. Lawrence (eds.), The Study of Time II, pp. 283-294. New York: Springer-Verlag.

Gunnel, John G. 1987.
Political Philosophy and Time: Plato and the Origins of Political Vision. Chicago, IL: University of Chicago Press.

Tucker, H.J. 1982.
"It's about time: the use of time in cross-sectional state policy research." American Journal of Political Science, 26:176-196.

Weinstein, Michael A. 1978.
Meaning and Appreciation: Time and Modern Political Life. West Lafayette, IN: Purdue University Press.

17

Psychiatry and Psychoanalysis

Abraham, Georges. 1976.
"The sense and concept of time in psychoanalysis." International Review of Psycho-Analysis, 3(4):461-472.

Arlow, Jacob A.. 1959.
"The structure of the deja vu experience." Journal of the American Psychoanalytic Association, 7:611-632.

Arlow, Jacob A. 1984.
"Disturbances of the sense of time -- with special reference to the experience of timelessness." Psychoanlaytic Quarterly, 53(1):13-37.

Arlow, Jacob A. 1986.
"Psychoanalysis and time." Journal of the American Psychoanalytic Association, 34(3):507-528.

Banks, Robin, Daniel Cappon, and Ross Hagen. 1966.
"Time estimation of psychiatric patients." Perceptual and Motor Skills, 23:1294.

Barber, Th. X., and D. Calverley. 1964.
"Toward a theory of 'hypnotic' behaviour: an experimental study of 'hypnotic time distortion'." Archives of General Psychiatry, 10(2):209.

Baum, Steve K., Russell L. Boxley, and Marcia Sokolowski. 1984.
"Time perception and psychological well-being in the elderly." Psychiatric Quarterly, 56(1):54-61.

Bergler, Edmund, and Geza Roheim. 1946.
"Psychology of time perception." Psychoanalytic Quarterly, 15:190-206.

Besier, Morton. 1987.
"Changing time perspective and mental health amoung Southeast Asian refugees." Culture, Medicine and Psychiatry, 11(4):437-464.

Bonaparte, Marie. 1940.
"Time and the unconscious." International Journal of Psycho-Analysis, 21:427-468.

Borbely, Antal F. 1987.
"Towards a temporal theory of the mind." Psychoanalysis and Contempory Thought, 10(4):459-487.

Boss, Medard. 1965.
Psychoanalysis and Daseinanalysis. New York: Basic Books.

Bromberg, W. 1938.
"The meaning of time for children." American Journal of Orthopsychiatry, 8:142-147.

Brotchie, Janet, James Brennan, and Maria A. Wyke. 1985.
"Temporal orientation in the pre-senium and old age." British Journal of Psychiatry, 147:692-695.

Butler, Robert N. 1963.
"The life review: an interpretation of reminiscence in the aged." Psychiatry, 26:65-76.

Calhoun, Craig Jackson. 1976.
"Continuity and change: the significance of time in the organization of experience." International Review of Psycho-Analysis, 3:291.

Cappon, Daniel, and Robin Banks. 1964.
"Experiments in time perception." Canadian Psychiatric Association Journal, 9(5):396-410.

Cohen, Louis H., and Gregory N. Rochlin. 1938.
"Loss of temporal localization as a manifestation of disturbed self-awareness." American Journal of Psychiatry, 95:87-95.

Cohen, Samuel I., and Alexander G. Mezey. 1961.
"The effect of anxiety on time judgment and time experience in normal persons." Journal of Neurology and Neurosurgical Psychiatry, 24(3):266-268.

Cohn, Franz S. 1957.
"Time and the ego." Psychoanalytic Quarterly, 26:168-189.

Colarusso, Calvin A. 1979.
"The development of time sense: from birth to object constancy." International Journal of Psycho-Analysis, 60:243-251.

Colarusso, Calvin A. 1987.
"The development of time sense: from object constancy to adolescence." Journal of the American Psychoanalytic Association, 35(1):119-144.

Cole, Malvin, and O.L. Zangwill. 1963.
"Deja vu and temporal lobe epilepsy." Journal of Neurology, Neurosurgery, and Psychiatry, 26:37.

Dahl, Max. 1958.
"A singular distortion of temporal orientation." American Journal of Psychiatry, 115:146-149.

Dapkus, Marilyn A. 1985.
"A thematic analysis of the experience of time." Journal of Personality and Social Psychology, 49(2):408-419.

Denber, H.C. 1986.
"The matter of time in psychiatry: a preliminary report." Psychiatric Journal of the University of Ottawa, 11(4):210-222.

Dooley, Lucile. 1941.
"The concept of time in defence of ego integrity." Psychiatry, 4:13-23.

Dubois, Franklin S. 1955.
"The sense of time and its relation to psychiatric illness." _American Journal of Psychiatry_, 111:46-51.

Ehrentheil, Otto F., and Peter B. Jenney. 1960.
"Does time stand still for some psychotics?" _Archives of General Psychiatry_, 3:1-3.

Eisenbud, Jule. 1956.
"Time and the oedipus". _Psychoanalytic Quarterly_, 25:363-384.

Eissler, K.R. 1952.
"Time experience and the mechanisms of isolation." _Psychoanalysis Review_, 39:1-21.

Elkin, D.G. 1963.
"Time perception and the feedback principle." _Soviet Psychology and Psychiatry_, 1(4):9-12.

Elkin, D.G. 1964.
"Time perception and anticipatory reflection." _Soviet Psychology and Psychiatry_, 3(3):42-48.

Fenchel, Gerd H. 1985.
"Time as self and ego experience." _Issues in Ego Psychology_, 8(1-2):73-81.

Frank, Robert G. 1984.
"Psychological response to amputation as a function of age and time since amputation." _British Journal of Psychiatry_, 144:493-497.

Frankl, V.E. 1966.
"Time and responsibility." _Existential Psychiatry_, 1:361-66.

Fraser, J.T. 1981.
"Temporal levels in reality testing." _International Journal of Psycho-Analysis_, 62:3-26.

Friedman, S., and C. Fisher. 1967.
"On the presence of a rhythmic, diurnal oral instinctual drive cycle in man." _Journal of the American Psychoanalytic Assocation_, 15:317-343.

Gifford, S. 1960.
"Sleep, time and the early ego: comments on the development of the 24-hour steep-wakefulness pattern as a precursor of ego functioning." Journal of the American Psychoanalytic Assocation, 8(1):5-42.

Goldberg, A. 1971.
"On waiting." International Journal of Psychoanalysis, 52:413-21.

Gross, A. 1949.
"Sense of time in dreams." Psychoanalytic Quarterly, 18:446-470.

Hartocollis, Peter. 1972.
"Time as a dimension of affects." Journal of the American Psychoanalytic Association, 20(1):92-108.

Hartocollis, Peter. 1974.
"Origins of time: a reconstruction of the ontogenetic development of the sense of time based on object-relations theory." Psychoanalytic Quarterly, 43:243-261.

Hartocollis, Peter. 1976.
"On the experience of time and its dynamics, with special reference to affects." Journal of the American Psychoanalytic Association, 24(2):363-375.

Hartocollis, Peter. 1978.
"Time and affects in borderline disorders." International Journal of Psycho-Analysis, 59:157-163.

Hartocollis, Peter. 1983.
Time and Timelessnes, or, The Varieties of Temporal Experience (A Psychoanalytic Inquiry). New York: International Universities Press.

Heuscher, Julius E. 1971.
"The temporal experience today: a phenomenological study." American Journal of Psychoanalysis, 31(2):192-204.

Ingram, Douglas H. 1979.
"Time and timekeeping in psychoanalysis and psychotherapy." American Journal of Psychoanalysis, 39(4):319-328.

Jenner, F.A., J.C. Goodwin, M. Sheridan, I.U. Tauber, and M.C. Lobban. 1968.
"The effects of altered time regime on biological rhythms in a 48-hour periodic psychosis." British Journal of Psychiatry, 111:215-224.

Kitamura, Toshinori, and Rajinder Kumar. 1983.
"Time estimation and time production in depressive patients." Acta Psychiatrica Scandinavica, 68(1):15-21.

Krippner, Stanley. 1969.
"The paranormal dream and man's pliable future." Psychoanalytic Review, 56(1):28-43.

Kubacki, Andrezj. 1987.
"Time and the meaning of human experience." American Journal of Psychiatry, 144(5):693-694.

Kurtz, Stephen A. 1988.
"The psychoanalysis of time." Journal of the American Psychoanalytic Association, 36(4):985-1004.

Lachmann, Frank M. 1985.
"On transience and the sense of temporal continuity." Contemporary Psychoanalysis, 21(2):193-200.

Leowald, Hans W. 1962.
"Superego and time." International Journal of Psycho-Analysis, 43:264-268.

Leowald, Hans W. 1972.
"The experience of time." The Psychoanalytic Study of the Child, 27:401-410.

Lomranz, Jacob, Dov Shmotkin, Amnon Zechovoy, and Eliot Rosenberg. 1985.
"Time orientation in Nazi concentration camp survivors: forty years after." American Journal of Orthopsychiatry, 55(2):230-236.

Lovell, K., and N. Slater. 1960.
"The growth of the concept of time: a comparative study." Journal of Child Psychology and Psychiatry, 1:179-190.

Lovett Doust, J. 1964.
"Biological clocks: psychophysiologic evidence for the existence of one such clock specific to certain

psychiatric disorders." Canadian Psychiatric Association Journal, 9(3):210.

Martinius, Joest. 1984.
"Complex visual time measurement under irregular and regular preparatory interval conditions in children with developmental language problems." Acta Paedopsychiatrica, 50(2-3):111-118.

Masler, E.G. 1973.
"The subjective perception of two aspects of time: duration and timelessness." International Journal of Psycho-Analysis, 54:425-429.

Meerloo, Joost A.M. 1948.
"Father time: an analysis of subjective conceptions of time." Psychiatric Quarterly, 22:587-608.

Meerloo, Joost A.M. 1950.
"Father time II." Psychiatric Quarterly, 24:657-671.

Meerloo, Joost A.M. 1981.
"The time sense in psychiatry." In J.T. Fraser (ed.), The Voices of Time, 2nd edn., pp. 235-252. Amherst, MA: University of Massachusetts Press.

Meisenhelder, Thomas M. 1985.
"An essay on time and the phenomenology of imprisonment." Deviant Behavior, 6(1):39-56.

Melges, F.T. 1972.
"Future oriented psychotherapy." American Journal of Psychotherapy, 26:22-23.

Melges, F.T. 1982.
Time and the Inner Future: A Temporal Approach to Psychiatric Disorders. New York: Wiley.

Melges, F.T., and C.E. Fongerousse. 1966.
"Time sense, emotions, and acute mental illness." Journal of Psychiatric Research, 4:127-140.

Melges, F.T., and A.M. Freeman. 1977.
"Temporal disorganization and inner-center confusion in acute mental illness." American Journal of Psychiatry, 132:874-877.

Melges, F.T., and A.E. Weisz. 1971.
"The personal future and suicidal ideation." Journal of Nervous and Mental Disease, 153:240-50.

Melges, F.T., et al. 1971.
"The personal future and self-esteem." Archives of General Psychiatry, 25:494-497.

Morris, James. 1983.
"Time experience and transference." Journal of the American Psychoanalytic Association, 31(3):651-675.

Mykel, Scott J. 1985.
"Discussion of Dr. G.H. Fenchel's paper 'Time as self and ego experience'." Issues in Ego Psychology, 8(1-2):82-86.

Oberndorf, C.P. 1941.
"Time: its relation to reality and purpose." Psychoanalytic Review, 28:139-155.

Orgel, S. 1965.
"On time and timelessness." Journal of the American Psychoanalytic Association, 13:102-121.

Orgel, S. 1976.
"On the experience of time and its dynamics, with special references to affects: dicussion. Journal of the American Psychoanalytic Association, 24(2):377-382.

Panel. 1972.
"The experience of time." Journal of the American Psychoanalytic Association, 20:650-667.

Peniazek, Z. 1982.
"The experience of time and hope in the elderly." Contemporary Psychoanalysis, 18(4):635-645.

Plessner, H. 1957.
"On the relation of time to death." In Joseph Campbell (ed.), Man and Time, pp. 233-263. New York: Panthoen.

Plunkett, James E., M. Schaefer, N. Kalter, and K. Okla. 1986.
"Perceptions of quality of life following divorce: a study of children's prognostic thinking." Psychiatry, 49(1):1-12.

Pollack, G.H. 1971.
"Temporal anniversary manifestation: hour, day, holiday." Psychoanalytic Quarterly, 40(1):123-131.

Pollack, G.H. 1971.
"On time, death, and immortality." Psychoanalytic Quarterly, 40:435-446.

Pollack, G.H. 1972.
"On time and anniversaries." In M. Kranzer (ed.), The Unconscious Today, pp. 233-257. New York: International Universities Press.

Pollack, I.W., et al. 1969.
"Social class and the subjective sense of time." Archives of General Psychiatry, 21(1):1-14.

Richter, C.P. 1965.
Biological Clocks in Medicine and Psychiatry, Sprinfield, IL: Charles C. Thomas.

Richter, C.P., and U. Benzehofer. 1985.
"Time estimation and chronopathology in endogenous depression." Acta Psychiatrica Scandinavica, 72(3):246-253.

Schiffer, Irvine. 1978.
The Trauma of Time: A Psychoanalytic Investigation. New York: International Universities Press.

Schimek, J.G. 1975.
"The interpretations of the past: childhood trauma, psychical reality, and historical truth." Journal of the American Psychoanalytic Association, 23:845-865.

Schlesinger, Jr., Arthur M. 1977.
"The modern consciousness and the winged chariot." In Bernard S. Gorman and Alden E. Wessman (eds.), The Personal Experience of Time, pp. 267-288. New York: Plenum.

Schlosberg, A. 1969.
"Time perspective in schizophrenia." Psychiatric Quarterly, 43:22-34.

Schneider, D.E. 1948.
"Time-space and growth of sense of reality." Psychoanalytic Review, 35(3):229-251.

Seeman, M.V. 1976.
"Time and schizophrenia." Psychiatry, 39:189-195.

Shectman, Fred. 1986.
"Time and the practice of psychotherapy." Psychotherapy, 23(4):521-525.

Smith, Henry F. 1988.
"Time, reconstruction and psychic reality." Journal of the American Academy of Psychoanalysis, 16(1):71-81.

Smith, Marian W. 1952.
"Different cultural concepts of past, present and future." Psychiatry, 15:395-400.

Spiegel, David. 1981.
"Man as timekeeper." American Journal of Psychoanalysis, 41(1):5-14.

Szajnberg, Nathan. 1985.
"Empty memory." Dynamic Psychotherapy, 3(2):171-181.

Terr, Lenore C. 1984.
"Time and trauma." Psychoanalytic Study of the Child, 39:633-665.

Thompson, D., G. Spivack, and M. Levine. 1960.
"The experience of time as a determinant of self-control." Archives of General Psychiatry, 3(5):563.

Torras de Bea, Eulalia, and Jose Rallo Romero. 1986.
"Past and present in interpretation." International Review of Psycho-Analysis, 13(3):309-321.

Yates, S. 1935.
"Some aspects of time difficulties and their relation to music." International Journal of Psycho-Analysis, 16:341-354.

18
Psychology

Aaronson, Bernard S. 1968.
"Hypnotic alterations of space and time." International Journal of Parapsychology, 10:6-31.

Aaronson, Bernard S. 1972.
"Behavior and the place names of time." In Henri Yaker, Humphry Osmond, and Frances Cheek (eds.), The Future of Time: Man's Temporal Environment, pp. 405-436. Garden City, NY: Doubleday.

Achamamba, B. 1987.
"Perceptions of time and I-E locus of control." Journal of the Indian Academy of Applied Psychology, 13(1):12-15.

Adam, N., B.S. Rosner, E.C. Hosick, and D.L. Clark. 1971.
"Effect of anesthetic drugs on time production and alpha rhythm." Perception and Psychophysics, 10:133-136.

Adams, J.A., and L.R. Creamer. 1962.
"Proprioception variables as determiners of anticipatory timing behavior." Human Factors, 4:217-222.

Adams, R.D. 1977.
"Intervening stimulus effects on category judgments of duration." Perception and Psychophysics, 21:527-534.

Agarwal, Adesh, and Manju Srivastava. 1981. "Significance of time perspective in reflection-impulsivity." Psychological Studies, 26(2):100-103.

Agarwal, Adesh, and Sushma Tiwari. 1988. "Future orientation: a mediator in temporal coding." International Journal of Psychology, 23(2):151-163.

Agarwal, Adesh, and Kaushal K. Tripathi. 1984. "Influence of prolonged deprivation, age and culture on the development of future orientation." European Journal of Social Psychology, 14(4):451-453.

Agarwal, Adesh, and Kaushal K. Tripathi. 1984. "Influence of prolonged deprivation, age and culture on the development of future orientation." European Journal of Social Psychology, Vol. 14(4) 451-453.

Agarwal, Adesh, Kaushal K. Tripathi, and Manju Srivastava. 1983. "Social roots and psychological implications of time perspective." International Journal of Psychology, 18(5):367-380.

Agarwal, Adesh, and L.B. Tripathi. 1978. "Time perspective: I. Theoretical considerations." Psychological Studies, 23(2):61-68.

Agarwal, Adesh, and L.B. Tripathi. 1979. "Time perspective: II. Development and empirical validation of a new tool." Psychological Studies, 24(1):59-65.

Agarwal, Adesh, and L.B. Tripathi. 1980. "Time perspective in achievement motivation." Psychologia: An International Journal of Psychology in the Orient, 23(1):50-62.

Ahmadi, Kate S. 1984. "Effects of social influences and waiting on time judgment." Perceptual and Motor Skills, 59(3):771-776.

Aiken, Jr., Lewis R. 1965. "Learning and retention in the estimation of short time intervals: a circuit and a study." Perceptual and Motor Skills, 20:509-517.

Akerboom, S., and Gert Ten Hoopen. 1983.
"The effect of a contralateral drone on the perceptual onset asynchrony of interaural tone sequences." Perception and Pschophysics, 33:571-574.

Akerboom, S., Gert Ten Hoopen, P. Olierook, and T. Van der Schaaf. 1983.
"Auditing spatial alternation transforms auditing time." Journal of Experimental Psychology: Human Perception and Performance, 9:882-897.

Albert, S. 1978.
"Time, memory, and affect: experimental studies of the subjective past." In J.T. Fraser, N. Lawrence, and D. Park (eds.), The Study of Time III, pp. 269-290. New York: Springer-Verlag.

Allan, Lorraine G. 1975.
"The relationship between the perception of successiveness and the perception of order." Perception and Psychophysics, 18:29-36.

Allan, Lorraine G. 1977.
"The time-order error in judgments of duration." Canadian Journal of Psychology, 31:24-31.

Allan, Lorraine G. 1979.
"The perception of time." Perception and Pychophysics, 26:340-354.

Allan. Lorraine G. 1983.
"Magnitude estimation of temporal intervals." Perception and Psychophysics, 33:29-42.

Allan, Lorraine G. 1984.
"Contingent aftereffects in duration judgements." Annals of the New York Academy of Sciences, 423:116-130.

Allan, Lorraine G., and Alfred B. Kristofferson. 1974.
"Successive discrimination: two models." Perception and Psychophysics, 15:37-46.

Allan, Lorraine G., and Alfred B, Kristofferson. 1974.
"Judgments about the duration of brief stimuli." Perception and Psychophysics, 15:434-440.

Allan, Lorraine G., and Alfred B. Kristofferson. 1974. "Psychophysical theories of duration discrimination." Perception and Psychophysics, 16:26-34.

Allan, Lorraine G., Alfred B. Kristofferson, and E.W. Wiens. 1971. "Duration discrimination of brief light flashes." Perception and Psychophysics, 9:327-334.

Allen, J.F. 1983. "Maintaining knowledge about temporal intervals." Communications of the Association for Computer Machinery, 26:832-843.

Allen, Sheila S., and Robert V. Heckel. 1978. "Death fantasies and future time perspective." Journal of Clinical Psychology, 34(2):419-420.

Allport, D.A. 1968. "Phenomenal simultaneity and the perceptual moment hypothesis." British Journal of Psychology, 59:395-406.

Ames, Louise Bates. 1946. "The development of the sense of time in the young child." Journal of Genetic Psychology, 68:97-125.

Amidon, A., and P. Carey. 1972. "Why five-year-olds cannot understand before and after." Journal of Verbal Learning and Verbal Behavior, 11:417-423.

Anast, Philip. 1965. "Temporal orientation: and adaptive process." Journal of Social Psychology, 65:33-39.

Arieti, Silvano. 1947. "The processes of expectation and anticipation." Journal of Nervous and Mental Disease. 106:471-481.

Arlin, Marshall. 1989. "The effects of physical work, mental work, and quantity on children's time perception." Perception and Psychophysics, 45(3):209-214.

Arlow, Jacob A. 1989. "Time as emotion." In J.T. Fraser (ed.), Time and Mind:

The Study of Time VI. Madison, CT: International Universities Press.

Aronson, Elliot, and Eugene Gerard. 1966.
"Beyond Parkinson's law: the effect of excess time on subsequent performance." Journal of Personality and Social Psychology, 3:336-339.

Aronson, Elliot, and David Landy. 1967.
"Further steps beyond Parkinson's law: a replication and extension of the excess time effect." Journal of Experimental Social Psychology, 3:274-285.

Ashton, R. 1976.
"Aspects of timing in child development." Child Development, 47:622-626.

Avant, L.L., P.J. Lyman, and J.R. Antes. 1975.
"Effects of stimulus familiarity upon judged visual duration." Perception and Psychophysics, 17:253-262.

Axel, Robert. 1924.
"Estimation of time." Archives of Psychology, 12, no. 74, pp. 1-77.

Axelrod, S., and L.T. Guzy. 1968.
"Underestimation of dichotic click rates: results using methods of absolute estimation and constant stimuli." Psychonomic Science, 12:133-134.

Bachorowski, Jo-Anne, and Joseph P. Newman. 1985.
"Impulsivity in adults: motor inhibition and time-interval estimation." Personality and Individual Differences, 6(1):133-136.

Back, Kurt W., Stephen R. Wilson, Morton D. Bogdonoff, and Willian G. Troyer. 1967.
"In-between times and experiemental stress." Journal of Personality, 35:456-473.

Back, Kurt W., and Linda Brookover Bourque. 1970.
"Life graphs: aging and cohort effects." Journal of Gerontology, 25:249-255.

Baer, Paul E., Don Charles Wukasch, and Sanford Goldstone. 1963.

"Time judgment and level of aspiration." Perceptual and Motor Skills, 16:648.

Baker, Curtis L., and Oliver J. Braddick. 1985. "Temporal properties of the short-range process in apparent motion." Perception, 14(2):181-192.

Baker, L. 1978. "Processing temporal relationship in single stories: effects of input sequence." Journal of Verbal Learning and Verbal Behavior, 17:559-572.

Baldwin Robert O., Donald H. Thor, and Dale E. Wright. 1966. "Sex differences in the sense of time: failure to replicate a 1904 study." Perceptual and Motor Skills, 22:398.

Balsam, Peter. 1984. "Relative time in trace conditioning." Annals of the New York Academy of Sciences, 423:221-227.

Banks, Robin, and Daniel Cappon. 1962. "Effect of reduced sensory output on time perception." Perceptual and Motor Skills, 14:74.

Barabasz, Arreed Franz. 1970. "Time estimation and temporal orientation in delinquents and non-deliquents: a re-examination." Journal of General Psychology, 82:265-267.

Barabasz, Arreed Franz. 1970. "Temporal orientation and academic acheivement in college." Journal of Social Psychology, 80:231-232.

Bardwick, Judith M. 1978. "Middle age and a sense of future." Merrill-Palmer Quarterly, 24(2):129-138.

Barndt, Robert J., and Donald M. Johnson. 1955. "Time orientation in delinquents." Journal of Abnormal and Social Psychology, 51:343-345.

Barton, Eva. 1986. "Adolescent views of the contemporary world: an international comparison." Educational Psychology in Practice, 2(2):45-48.

Bascue, L.O., and R.E. Lawrence. 1977.
"A study of subjective time and death anxiety in the elderly." Journal of Death and Dying, 8(1):81-90.

Bassili, J. 1976.
"Temporal and spatial contiguity in the perception of social events." Journal of Personality and Social Psychology, 33:680-685.

Bauer, Patricia J., and Jean M. Mandler. 1989.
"One thing follows another: effects of temporal structure on 1- to 2-year-olds' recall of events." Developmental Psychology, 25(2):197-206.

Bauer, Rudolph, and John Gillies. 1972.
"Cognitive style and influence of success and failure on future time-perspective." Perceptual and Motor Skills, 34:79-82.

Bauer, Rudolph, and John Gillies. 1972.
"Measure of affective dimensions of future time perspective." Perceptual and Motor Skills, 34(1):181-182.

Baugh, Bernard T., and Susan E. Mason. 1986.
"Need for cognition related to time perception." Perception and Motor Skills, 62(2):540-542.

Beg, Moazziz A. 1985.
"Precognition, awareness and the space-time metrical field." Journal of Indian Psychology, 4(2):98-103.

Beilin, Harry. 1956.
"The pattern of postponability and its relation to social class mobility." Journal of Social Psychology, 44:33-48.

Bell, C.R. 1965.
"Time estimation and increases in body temperature." Journal of Experimental Psychology, 70:232-234.

Bell, C.R., and Anne N. Watts. 1966.
"Personality and judgments of temporal intervals." British Journal of Psychology, 57:155-159.

Ben Baruch, Ephraim, James E. Bruno, and Loretta L. Horn. 1985.

"Development of time horizon perceptions in students." Journal of the Association for the Study of Perception, 20(2):5-16.

Ben Baruch, Ephraim, James E. Bruno, and L. Lolly Horn. 1987. "Dimension of time use attitudes among middle-high SES students." Social Behavior and Personality, 15(1):1-12.

Bentley, Alastair M. 1983. "Personal and global futurity in Scottish and Swazi students." Journal of Social Psychology, 121(2):223-229.

Bentley, Alastair M. 1987. "Swazi children's understanding of time concepts: a Piagetian study." Journal of Genetic Psychology, 148(4):443-452.

Benton, A.L., M.W. Van Allen, and M.L. Fogel. 1964. "Temporal orientation in cerebral disease." Journal of Nervous and Mental Disease, 139:110-119.

Berman, A. 1939. "The relation of time estimation to satiation. " Journal of Experimental Psychology, 25:281-293.

Berndt, T.J., and D.J. Wood. 1974. "The development of time concepts through conflict based on a primitive duration capacity." Child Development, 45:825-828.

Bezak, Josef, and Stanislav Dornic. 1968. "Effect of a sensori-motor activity on the estimation of movement velocity." Studia Psychologica, 10:73-74.

Berzonsky, M. 1971. "The role of familiarity in children's explanations of physical causality." Child Development, 42:705-712.

Bilting, Marianne, et al. 1985. "Estimation of time as a measure of pain magnitude." Journal of Psychosomatic Research, 27(6):493-497.

Bindra, Dalbir, and Helene Waksberg. 1956. "Methods and terminology in studies of time estimation." Psychological Bulletin, 53:155-159.

Bingham, Dianne, and B. Jo Hailey. 1989.
"The time-urgency component of the Type A behavior pattern: time pressure and performance." Journal of Applied Social Psychology, 19(5):425-432.

Bjork, R.A. 1972.
"Theoretical implications of directed forgetting." In A.W. Melton and E. Martin (eds.), Coding Processes in Human Memory, pp. 217-235. Washington, DC: Winston.

Bjorkman, Mats. 1984.
"Decision making, risk taking and psychological time: review of empirical findings and psychological theory." Scandinavian Journal of Psychology, 25(1):31-49.

Blanchard, Harry E. 1985.
"A comparison of some processing time measures based on eye movements." Acta Psychologica, 58(1):1-15.

Block, Richard A. 1974.
"Memory and the experience of duration in retrospect." Memory and Cognition, 2:153-160.

Block, Richard A. 1978.
"Remembered duration: effects of event and sequence complexity." Memory and Cognition, 6:320-326.

Block, R.A. 1979.
"Time and consciousness." In G. Underwood and R. Stevens (eds.), Aspects of Consciousness, Vol. 1: Psychological Issues, pp. 179-217. London, UK: Academic.

Block, Richard A. 1982.
"Temporal judgments and contextual change." Journal of Experimental Psychology: Learning, Memory, and Cognition, 8:530-544.

Block, Richard A. 1985.
"Contextual coding in memory: studies in remembered duration." In John A. Michon and Janet L. Jackson (eds.), Time, Mind, and Behavior, pp. 169-178. New York: Springer-Verlag.

Block, Richard A. 1989.
"A contextualist view of time and mind." In J.T. Fraser (ed.), Time and Mind: The Study of Time VI. Madison, CT: International Universities Press.

Block, Richard A., E.G. George, and M.A. Reed. 1980. "A watched pot sometimes boils: a study of duration experience." _Acta Psychologica_, 46:81-94.

Block, Richard, and M.A. Reed. 1978. "Remembered duration: evidence for a contextual-change hypothesis." _Journal of Experimental Psychology: Human Learning and Memory_, 4:656-665.

Block, Richard A., John L. Saggau, and Leo H. Nickol. 1983-84. "Temporal inventory on meaning and experience: a structure of time." _Imagination, Cognition and Personality_, 3(3):203-225.

Boardman, W.K., et al. 1963. "Sociometric status and apparent duration." _Journal of Social Psychology_, 61(2):303-310.

Bochner, S., and K.H. David. 1968. "Delay of gratification, age and intelligence in an aboriginal culture." _International Journal of Psychology_, 3(3):167-174.

Bokander, Ingvar. 1965. "Time estimation as an indicator of attention-arousal when perceiving complex and meaningful stimulus material." _Percept and Motor Skills_, 21:323-328.

Boltz, Marilyn, Elizabeth Marshburn, Mari Riess Jones, and Walter W. Johnson. 1985. "Serial-pattern structure and temporal-order recognition." _Perception and Psychophysics_, 37(3):209-217.

Bond, M.J., and N.T. Feather. 1988. "Some correlates of structure and purpose in the use of time." _Journal of Personality and Social Psychology_, 55(2):321-329.

Boniecki, George J. 1977. "Is man interested in his future?" _International Journal of Psychology_, 12(1):59-64.

Boniecki, George J. 1978. "Man's concern for his future still in doubt." _International Journal of Psychology_, 13:239-244.

Boring, Edwin G. 1936.
"Temporal experiences and operationism." American Journal of Psychology, 48:519-522.

Bortner, Raymond W., and David F. Hultsch. 1972.
"Personal time perspective in adulthood." Developmental Psychology, 7(2):98-103.

Bottger, Preston C. 1984.
"Expertise and air time as bases of actual and perceived influence in problem-solving groups." Journal of Applied Psychology, 69(2):214-221.

Bouffard, Leandre, Willy Lens, and Joseph R. Nuttin. 1983.
"Extension of future time perspective in relation to frustration." International Journal of Psychology, 18(5):429-442.

Boulter, Lawrence R., and Mortimer H. Appley. 1967.
"Time and effort as determiners of time-production error." Journal of Experimental Psychology, 75:447-452.

Bowden, E.A.F. 1972.
"Accuracy of time estimation among rural and urban Buganda." Perceptual and Motor Skills, 28:54.

Bowden, Edward M. 1985.
"Accessing relevant information during problem solving: time constraints on search in the problem space." Memory and Cognition, 13(3):280-286.

Boynton, R.M. 1961.
"Some temporal factors in vision." In W.A. Rosenblith (ed.), Sensory Communication, pp. 739-756. New York: Wiley.

Braddeley, A.D. 1966.
"Time-estimation at reduced body temperature." American Journal of Psychology, 79:475-479.

Bradley, N.C. 1947.
"The growth of the knowledge of time in children of school age." British Journal of Psychology, 38:67-78.

Brannigan, Gary G., and Alexander Tolor. 1971.
"Sex differences in adaptive styles." Journal of Genetic Psychology, 119(1):143-149.

Brannigan, Gary G., and Alexander Tolor. 1971.
"Self-parental distance, control of perspective, and personal future time perspective." Journal of Genetic Psychology, 119(1):151-157.

Bray, D.H. 1970
"Extent of future time orientation: a cross-ethnic study among New Zealand adolescents." British Journal of Educational Psychology, 40(2):200-208.

Brigner, Willard L. 1984.
"Rotation of space-time plane predicts a new illusion of spatial displacement." Perceptual and Motor Skills, 59(2):359-369.

Brock, Timothy C., and Carolyn Del Guidice. 1963.
"Stealing and temporal orientation." Journal of Abnormal and Social Psychology, 66:91-94.

Bronckart, J.P., and H. Sinclair. 1973.
"Time, tense, aspect." Cognition, 2:107-130.

Brooks, L.R. 1968.
"Spatial and verbal components of the act of recall." Canadian Journal of Psychology, 22:349-368.

Brown, D.R., and Lloyd Hitchcock, Jr. 1965.
"Time estimation: dependence and independence of modality-specific effects." Perceptional and Motor Skills, 21:727-734.

Brown, L.B. 1965.
"Religious beliefs and judgment of brief duration." Perceptual and Motor Skills, 20:33-34.

Brown, Norman R., Lance J. Rips, and Steven K. Shevell. 1985.
"The subjective dates of natural events in very-long-term memory." Cognitive Psychology, 17(2):139-177.

Brown, Scott W. 1984.
"Retrospective duration judgments of a hypnotic time interval." Annals of the New York Academy of Sciences, 423:583-584.

Brown, Scott W. 1985.
"Time perception and attention: the effects of

prospective versus retrospective paradigms and task demands on perceived duration." Perception and Psychophysics, 38(2):115-124.

Buck, John N. 1946.
"The time appreciation test." Journal of Applied Psychology, 30:388-398.

Buhler, Charlotte. 1968.
"The course of human life as a psychological problem." Human Development, 11:184-200.

Bullock, Merry, and Rochel Gelman. 1979.
"Children's assumptions about cause and effect: temporal ordering." Child Development, 50:89-96.

Burd, Larry, John M. Dodd, and Wayne W. Fisher. 1984.
"Making functional time estimates: comparison of school-aged, learning-disabled and other children." Perceptual and Motor Skills, 59(3):967-973.

Burger, Jerry M., and Rose M. Huntzinger. 1985.
"Temporal effects on attributions for one's own behavior: the role of task outcome." Journal of Experimental Social Psychology, 21(3):247-261.

Burke, Ronald J. 1985.
"Beliefs and fears underlying Typr A behavior: correlates of time urgency and hostility." Journal of General Psychology, 112(2):133-145.

Burns, Neal M., and E.C. Gifford. 1961.
"Time estimation and anxiety." Journal of Psychological Studies, 12:19-27.

Burton, Arthur. 1943.
"A further study of the relation of time estimation to monotony." Journal of Applied Psychology, 27:350-359.

Butters, Nelson, Evelyn Jones, Joseph Hoyle, and Caroline Zsambok. 1968.
"Effect of dynamic verbal and tonal stimuli on the perception of time." Perceptual and Motor Skills, 27:431-437.

Cahoon, Richard L. 1967.
"Effect of acute exposure to altitude on time estimation." Journal of Psychology, 66:321-324.

Calabresi, R., and John Cohen. 1968.
"Personality and time attitudes." Journal of Abnormal Psychology, 73:431-439.

Cameron, Paul, K.G. Desai, Darius Bahador, and G. Dremel. 1977-78.
"Temporality across the life-span." International Journal of Aging and Human Development, 8(3):229-259.

Campagnoni, F.R., and P.S. Cohen. 1984.
"Scalar timing and the spatial organization of behavior between reward presentations." Annals of the New York Academy of Sciences, 423:585-587.

Campos, Leonard P. 1966.
"Relationship between time estimation and retentive personality traits." Perceptional and Motor Skills, 23:59-62.

Cappon, Daniel, and Milo Tyndel. 1967.
"Time perception in alcoholism." Quarterly Journal of Studies on Alcohol, 28:430-435.

Carlson, V.R., and I. Feinberg. 1968.
"Individual variation in time judgment and concept of an internal clock." Journal of Experimental Psychology, 77(4):631-640.

Carlson, V.R., and I. Feinberg. 1970.
Time judgment as a function of method, practice, and sex." Journal of Experimental Psychology, 85:171-180.

Carter, David E., and Glenn J. MacGrady. 1966.
"Acquisition of a temporal discrimination by human subjects." Psychonomic Science, 5:309-310.

Catania, A.C. 1970.
"Reinforcement schedules and psychophysical judgment: a study of some temporal properties of behavior." In W.N. Schoenfield (ed.), The Theory of Reinforcement Schedules, pp. 1-42. New York: Appleton-Century-Crofts.

Chatterjea, Ram. G. 1960.
"The effect of distraction on time gap in the estimation of short duration." British Journal of Psychology, 51(4):335-339.

Chatterjea, Ram. G. 1960.
"Time gap in the estimations of short duration: visual presentation." Indian Journal of Psychology, 35:147-158.

Chatterjea, Ram G. 1964.
"Temporal duration: ratio scale and category scale." Journal of Experimental Psychology, 67:412-416.

Chatterjea, Ram G., and Purabi Rakshit. 1966.
"Estimation of temporal interval." Perceptual and Motor Skills, 22:176.

Cheek, Frances E. 1972.
"The family context of the time perspective of schizophrenics." In Henri Yaker, Humphry Osmond, and Frances Cheek (eds.), The Future of Time: Man's Temporal Environment, pp. 310-329. Garden City, NY: Doubleday.

Cheek, Frances E., and Joan Laucius. 1972.
"The time world of three drug-using groups -- alcoholics, heroin addicts, and psychedelics." In Henri Yaker, Humphry Osmond, and Frances Cheek (eds.), The Future of Time: Man's Temporal Environment, pp. 310-329. Garden City, NY: Doubleday.

Chiriboga, David A. 1978.
"Evaluated time: a life course perspective." Journal of Gerontology, 33(3):388-393.

Christie, L.S., and R. Duncan Luce. 1956.
"Decision structure and time relations in simple choice behavior." Journal of Mathematical Psychology, 18:89-112.

Church, Russell M. 1980.
"Short-term memory for time intervals." Learning and Motivation, 11:208-219.

Clark, Herbert. 1973.
"Space, time, semantics, and the child." In Timothy E. Moore (ed.), Cognitive Development and the Acquisition of Language, pp. 27-63. New York: Academic.

Clarke, E. 1982.
"Timing in the performance of Erik Satie's 'Vexations'." *Acta Psychologica*, 50:1-19.

Clausen, Johs. 1950.
"An evaluation of experimental methods of time judgment." *Journal of Experimental Psychology*, 40:756-761.

Clements, William M. 1979.
"The sense of life time in human development." *Journal of Religion and Health*, 18(2):88-92.

Cloudsley-Thompson, J.L. 1981.
"Time sense of animals." In J.T. Fraser (ed.), *The Voices of Time*, 2nd edn., pp. 296-311. Amherst, MA: University of Massachusetts Press.

Cohen, Gillian. 1977.
The Psychology of Cognition. London, UK: Academic Press.

Cohen, John. 1954.
"The experience of time." *Acta Psychologica*, 10:207-219.

Cohen, John. 1964.
"Psychological time." *Scientific American*, 211(5):116-124.

Cohen, John. 1967.
Psychological Time in Health and Disease. Springfield, IL: Charles C. Thomas.

Cohen, John. 1971.
"Time in psychology." In Jiri Zeman (ed.), *Time in Science and Philosophy*, pp. 153-164. New York: Elsevier.

Cohen, John. 1981.
"Subjective time." In J.T. Fraser (ed.), *The Voices of Time*, 2nd edn., pp. 257-275. Amherst, MA: University of Massachusetts Press.

Cohen, John, and Peter Cooper. 1962.
"New phenomena in apparent duration, distance and speed." *Nature*, 196:1233-1234.

Cohen, John, Peter Cooper, and Akio Ono. 1963.
"The hare and tortoise: a study of the *tau*-effect in walking and running. *Acta Psychologica*, 21:387-393.

Cohen, John, Peter Cooper, and G.B. Skelly. 1966. "Interrelations between judgements made by drivers, of the duration, distance and speed of their journeys." *Operational Research Quarterly*, 17(3):279-289.

Cohen, John, C.E.M. Hansel, and J.D. Sylvester. 1954. "An experimental study of comparative judgments of time." *British Journal of Psychology*, 45(2):108-114.

Cohen, John, C.E.M. Hansel, and J.D. Sylvester. 1955. "Interdependence in judgments of space, time and movement." *Acta Psychologica*, 11:360-372.

Cohen, Stewart, and Oscar A. Parsons. 1964. "The perception of time in patients with coronary artery disease." *Journal of Psychosomatic Research*, 8:1-7.

Collard, R., and Dirk-Jan Povel. 1982. "Theory of serial pattern production: tree traversals." *Psychological Review*, 89:693-707.

Collard, R.F.A., and E.L.J. Leeuwenberg. 1981. "Temporal order and spatial context." *Canadian Journal of Psychology*, 35:323-329.

Collard, R.R.A., and H.F.J.M. Buffart. 1982. "Minimization of structural information: a set-theoretical approach." *Pattern Recognition*, 6:21-32.

Colquhoun, W.P. (ed.). 1971. *Biological Rhythm and Human Performance*. New York: Academic.

Coltheart, Max, and Guy Von Sturmer. 1968. "A serial effect in time estimation." *Psychonomic Science*, 10:283-284.

Conroy, R., and J.N. Mills. 1970. *Human Circadian Rhythms*. London, UK: Churchill.

Cooper, Linn F., and Milton D. Erickson. 1959. *Time Distortion in Hypnosis*. Baltmore, MD: Williams and Wilkins.

Cooper, Pamela E., L. Eugene Thomas, Scott J. Stevens, and David Suscovich. 1981.

"Subjective time experience in an intergenerational sample." _International Journal of Aging and Human Development_, 13:183-193.

Copple, C., and R. Coon. 1977.
"The role of causality in encoding and remembering events as a function of age." _Journal of Genetic Psychology_, 130:129-136.

Corwin, T.R., and R.M. Boynton. 1968.
"Transitivity of visual judgments of simultaneity." _Journal of Experimental Psychology_, 78:560-568.

Costa, Paul, and Robert J. Kastenbaum. 1967.
"Some aspects of memories and ambitions in centenarians." _Journal of Genetic Psychology_, 110(1):3-16.

Cottle, Thomas J. 1967.
"The circles test: an investigation of temporal relatedness and dominance." _Journal of Projective Techniques and Personality Assessment_, 31(5):58-71.

Cottle, Thomas J. 1968.
"The location of experience: a manifest time orientation." _Acta Psychologica_, 28:129-149.

Cottle, Thomas J. 1969.
"The duration inventory: subjective extensions of temporal zones." _Acta Psychologica_, 29:333-352.

Cottle, Thomas J. 1969.
"Temporal correlates of the achievement value and manifest anxiety." _Journal of Consulting and Clinical Psychology_, 33:541-50.

Cottle, Thomas J. 1976.
Perceiving Time: A Psychological Investigation with Men and Women. New York: Wiley.

Cottle, Thomas J. 1977.
"The time of youth." In Bernard S. Gorman and Alden E. Wessman (eds.), _The Personal Experience of Time_, 163-189. New York: Plenum.

Cottle, Thomas J., and Peter Howard. 1969.
"Temporal extension and time zone bracketing in Indian adolescents." _Perceptual and Motor Skills_, 28:599-612.

Cottle, Thomas J., and Peter Howard. 1971. "Temporal correlates of dogmatism." Journal of Consulting and Clinical Psychology, 36(1):70-81.

Cottle, Thomas J., Peter Howard, and Joseph H. Pleck. 1969. "Adolescent perception of time: the effect of age, sex, and social class." Journal of Personality, 37:636-650.

Cottle, Thomas J., Peter Howard, and Joseph H. Pleck. 1969. "Linear estimations of temporal extension: the effect of age, sex, and social class." Journal of Projective Techniques and Personality Assessment, 33:81-93.

Cottle, Thomas J., and Stephen L. Klineberg. 1974. The Present of Things Future: Explorations of Time in Human Experience. New York: Free Press.

Cottle, Thomas J., Joseph Pleck, and Sudhir Kakar. 1968. "Time and content of significant life experiences." Perceptual and Motor Skills, 27:155-171.

Craig, Angus, and Ruth Condon. 1984. "Operational efficiency and time of day." Human Factors, 26(2):197-205.

Craig, Angus, and Ruth Condon. 1985. "Speed-accuracy trade-off and time of day." Acta Psychologica, 58(2):115-122.

Craik, F.I.M., and R.S. Lockhart. 1972 "Levels of processing: a framework for memory research." Journal of Verbal Learning and Verbal Behavior, 11:671-684.

Craik, Kenneth H., and Theodore R. Sabin. 1963. "Effect of covert alterations of clock rate upon time estimations and personal tempo." Perceptual and Motor Skills, 16:597-610.

Crawford, M.L.J., and Donald H. Thor. 1967. "Time perception in children in the absence of external temporal synchronizers." Acta Psychologica, 26:182-188.

Cromer, Gerald. 1988. "Time perspectives and time usage in a basic training course in the Israeli Defense Forces." Small Group Behavior, 19(1):67-78.

Cromer, R. 1971.
"The development of the ability to decenter in time." British Journal of Psychology, 62:353-365

Cromwell, Ruel L., David Rosenthal, David Shakow, and Theodore R. Kahn. 1961.
"Reaction time, locus of control, choice behavior, and descriptions of parental behavior in schizophrenic and normal subjects." Journal of Personality, 29:363-380.

Crowder, A.M., and R.H. Hohle. 1970.
"Time estimation by young children with and without informational feedback." Journal of Experimental Child Psychology, 10:353-365.

Cummings, L.L. 1982.
"Organizational behavior." Annual Review of Psychology, 33:541-579.

Cunningham, T.F., A.F. Healy, and D.M. Williams. 1984.
"Effects of repetition on short-term retention of order information." Journal of Experimental Psychology: Learning, Memory and Cognition, 10:575-597.

Curtis, J.S. 1916.
"Duration and temporal judgment." American Journal of Psychology, 27:1-46.

Daltrey, Marilyn H., and Philip Langer. 1984.
"Development and evaluation of a measure of future time perspective." Perceptual and Motor Skills, 58(3):719-725.

D'Amato, M.R. 1973.
"Delayed matching and short-tem memory in monkeys." In G.H. Bower (ed.), The Psychology of Learning and Motivation, Vol. 7, pp. 227-269. New York: Academic.

Dannenbring, G.L., and A.S. Bregman. 1976.
"Stream segregation and the illusion of overlap." Journal of Experimental Psychology: Human Perception and Performance, 2:544-555.

Danziger, K. 1965.
"Effect of variable stimulus intensity on estimates of duration." Perceptual and Motor Skills, 20:505-508.

Danziger, K., and Peter D. Du Preez. 1963.
"Reliability of time estimation by the method of reproduction." Perceptual and Motor Skills, 16:879-884.

Dapkus, Marilyn A. 1985.
"A thematic analysis of the experience of time." Journal of Personality and Social Psychology, 49(2):407-419.

Das, J.P. 1983.
"Memory for spatial and temporal order in deaf children." American Annals of the Deaf, 128(7):894-899.

Davids, Anthony, Catherine Kidder, and Melvyn Reich. 1962.
"Time orientation in male and female delinquents." Journal of Abnormal and Social Psychology, 64:237-240.

Davids, Anthony, and Anita Negrin Parenti. 1958.
"Time orientation and interpersonal relations of emotionally disturbed and normal children." Journal of Abnormal and Social Psychology, 57:299-305.

Davids, Anthony, and Jack Sidman. 1962.
"A pilot study -- impulsivity, time orientation, delayed gratification in future scientists and in underachieving high school students." Exceptional Children, 29:170-174.

Day, H.I. 1968.
"Some determinants of looking time under different instructional sets." Perception and Psychophysics, 4:279-281.

DeCasper, A.J., and M.D. Zeiler. 1974.
"Time limits for completing fixed ratios: III. Stimulus variables." Journal of the Experimental Analysis of Behavior, 22:275-283.

DeCasper, A.J., and M.D. Zeiler. 1977.
"Time limits for completing fixed ratios: IV. Components of the ratio." Journal of the Experimental Analysis of Behavior, 22:285-300.

De Charms, R. 1968.
Personal Causation. New York: Academic.

Delay, Eugene R., and Mianne E. Mathey. 1985.
"Effects of ambient noise on time estimation by humans." Perceptual and Motor Skills, 61(2):415-419.

De Man, Anton F., Gladys M. Batten, Christopher D. Green, and Lawrence Weinstein. 1983.
"Birth order and past-orientation." Psychological Reports, 53(2):425-426.

Demany, L., B. McKenzie, and E. Vurpillot. 1977.
"Rhythm perception in early infancy." Nature, 266:718-719.

Dempsey. A. 1971.
"Time conservation across cultures." International Journal of Psychology, 6:115-120.

Deregowski, Jan B. 1970.
"Effect of cultural value of time upon recall." British Journal of Social and Clinical Psychology, 9:115-120.

Deutsch, Diana. 1979.
"Binaural integration of melodic pattern." Perception and Psychophysics, 25:399-405.

Deutsch, Diana. 1980.
"The processing of structured and unstructured tonal sequences." Perception and Psychophysics, 28:381-389.

Deutsch, Diana. 1986.
"Recognition of durations embedded in temporal patterns." Perception and Psychophysics, 39(3):179-186.

Devisch, Renaat. 1985.
"Approaches to symbol and symptom in bodily space-time." International Journal of Psychology, 20(4-5):389-415.

De Volder, Maurice L. 1979.
"Time orientation: a review." Psychologica Belgica, 19(1):61-79.

De Volder, Maurice L., and Willy Lens. 1982.
"Academic achievement and future time perspective as a cognitive-motivational concept." Journal of Personality and Social Psychology, 42(3):566-571.

De Wolfe, Ruthanne K.S., and Carl P. Duncan. 1959.
"Time estimation as a function of level of behavior of successive tasks." Journal of Experimental Psychology, 58:153-158.

Dickie, Jane R., Thomas E. Ludwig, and David Blauw. 1979. "Life satisfaction among institutionalized and non-institutionalized older adults." Psychological Reports, 44(3,Pt.1):807-810.

Dickstein, L.S., and S.J. Blatt. 1966. "Death concern, futurity and anticipation." Journal of Consulting Psychology, 30:11-17.

Dilling, Carole A., and Albert I. Rabin. 1967. "Temporal experience in depressive states and schizophrenia." Journal of Consulting Psychology, 31:604-608.

Dimond, Stuart J. 1964. "The structural basis of timing." Psychological Bulletin, 62:348-350.

Dmitriev, A.S., and A.M. Kochigina. 1959. "The importance of time as a stimulus of conditioned reflex activity." Psychological Bulletin, 56:106-132.

Dobbs, H.A.C. 1951. "The relation between the time of psychology and the time of physics - Part I." British Journal for the Philosophy of Science, 2:122-141.

Dobbs, H.A.C. 1965. "Time and extrasensory perception." Proceedings of the Society for Psychical Research, 54:249-361.

Dobson, William R. 1954. "An investigation of various factors involved in time perception as manifested by different nosological groups." Journal of General Psychology, 50:277-298.

Dodd, John M., Philip E. Griswold, Gilbert H. Smith, and Larry Burd. 1985. "A comparison of learning disabled and other children on the ability to make functional time estimates." Child Study Journal, 15(3):189-197.

Doehring, D.G. 1961. "Accuracy and consistency of time-estimation by four methods of reproduction." American Journal of Psychology, 74:27-35.

Dosher, Barbara A. 1984.
"Degree of learning and retriveal speed: study time and multiple exposures." Journal of Experimental Psychology: Learining, Memory, and Cognition, 10(4):541-574.

Dowling, D.J. 1973.
"The perception of interleaved melodies." Cognitive Psychology, 5:322-337.

Dowling, D.J. 1978.
"Scale and contour: two components of a theory of memory for melodies." Psychological Review, 85:341-354.

Drewnowski, A. 1980.
"Attributes and priorites in short-term recall: a new model of memory span." Journal of Experimental Psychology: General, 109:208-250.

Ducasse, C.J. 1976.
"A theory of the relation of causality to recognition." In James M.O. Wheatley and Hoyt L. Edge (eds.), Philosophical Dimensions of Parapsychology, pp. 227-234. Springfield, IL: Charles C. Thomas.

Du Preez, Peter D. 1964.
"Judgment of time and aspects of personality." Journal of Abnormal and Social Psychology, 69:228-233.

Du Preez, Peter D. 1967.
"Field dependence and accuracy of comparison of time intervals." Perceptual and Motor Skills, 24:467-472.

Du Preez, Peter D. 1967.
"Reproduction of time intervals after short periods of delay." Journal of General Psychology, 76:59-71.

Dziurawiec, S., and Jan B. Deregowski. 1986.
"Time as a factor in a spatial task." International Journal of Psychology, 21(2):177-187.

Edlund, Mathew. 1987.
Psychological Time and Mental Illness. New York: Gardner Press.

Edmunston, Jr., William E., and John R. Erbeck. 1967.
"Hypnotic time distortion: a note." American Journal of Clinical Hypnosis, 10:79-80.

Efron, R. 1963.
"The effect of handedness on the perception of simultaneity and temporal order." Brain, 86:261-284.

Efron, R. 1963.
"The effect of stimulus intensity on the perception of simultaneity of right and left-handed subjects." Brain, 86:285-294.

Efron, R. 1963.
"Temporal perception, aphasia, and deja vu." Brain, 86:403-424.

Efron, R. 1967.
"The duration of the present." Annals of the New York Academy of Sciences, 138:713-729.

Efron, Robert. 1972.
"The measurement of perceptual durations." In J.T.Fraser, F.C. Haber, and G.H. Muller (eds.), The Study of Time, pp. 257-268. New York: Springer-Verlag.

Efron, Robert. 1973.
"An invariant characteristic of perceptual systems in the time domain." In S. Kornblum (ed.), Attention and Performance IV, pp.713-736. New York: Academic Press.

Ehrlich, S. 1960.
"Rhythm: recent French contributions." Acta Psychologica, 17:155-176.

Eisler, H. 1975.
"Subjective duration and psychophysics." Psychological Review, 82:429-450.

Eisler, H. 1976.
"Experiments on subjective duration 1868-1975: a collection of power function experiments." Psychological Bulletin, 83:1154-1171.

Eisler, H. 1981.
"The parallel-clock model: replies to critics and criticisms." Perception and Psychophysics, 29:516-520.

Eizzler, Charles F. 1983.
"Perceptual preferences as an aspect of adolescent learning styles." Education, 103(3):231-242.

Ekman, Gosta, and Ulf Lundberg. 1971.
"Emotional reaction to past and future events as a function of temporal distance." _Acta Psychologica_, 35(6):430-441.

Ellerman, H., and John A. Michon. 1983.
"The psychophysics of time reproduction." In C. Micko and U. Schulz (eds.), _Formalization of Psychological Theories_, pp. 38-50. University of Bielfeld.

Elliott, Didby, and Ruth Jones. 1984.
"Short-term memory for duration of movement." _Perceptual and Motor Skills_, 58(1):331-334.

El-Meligi, A. Moneim. 1972.
"A technique for exploring time experiences in mental disorders." In Henri Yaker, Humphry Osmond, and Frances Cheek (eds.), _The Future of Time: Man's Temporal Environment_, pp. 220-271. Garden City, NY: Doubleday.

Emley, G.S., C.R. Schuster, and B.R. Lucchesi. 1968.
"Trends observed in the time estimation of three stimulus intervals within and across sessions." _Perceptual and Motor Skills_, 26:391-398.

Epley, D., and D. Ricks. 1963.
"Foresight and hindsight in the TAT." _Journal of Projective Techniques_, 27:51-59.

Erickson, Milton D., and Elizabeth M. Erickson. 1958.
"Further considerations of time distortion: subjective time condensation as distict from time expansion." _American Journal of Clinical Hypnosis_, 1:83-88.

Eson, Morris E., and John S. Kafka. 1952.
"Diagnostic implications of a study in time perception." _Journal of General Psychology_, 46:169-183.

Eson, Morris E., and Norman Greenfield. 1962.
"Lifespace: its contents and temporal dimensions." _Journal of Genetic Psychology_, 100:113-128.

Essens, Peter J. 1986.
"Hierarchical organization of temporal patterns." _Perception and Psychophysics_, 40(2):69-73.

Essens, Peter J. and Dirk-Jan Povel. 1985.
"Metrical and nonmetrical representations of temporal patterns." Perception and Psychophysics, 37(1):1-7.

Estes, William K. 1972.
"An associative basis for coding and organization in memory." In A.W. Melton and E. Martin (eds.), Coding Processes in Human Memory, pp. 161-190. Washington, DC: V.H. Winston.

Estes, William K. 1980.
"Is human memory obsolete?" American Scientist, 68:62-69.

Estes, William K. 1982.
"Multiple coding and processing stages: a review." In F. Klix, J. Hofmann, and E. Van der Meer (eds.), Cognitive Research in Psychology, pp. 14-21. Berlin, FRG: VEB Deutscher Verlag der Wissenschaften.

Estes, William K. 1985.
"Memory for temporal information." In John A. Michon and Janet L. Jackson (eds.), Time, Mind, and Behavior, pp. 151-168. New York: Springer-Verlag.

Etzioni, Amitai, and R. Remp. 1972.
"Life space: its content and temporal dimensions." Journal of Genetic Psychology, 100, pp. 113-28.

Ezekiel, Raphael S. 1968.
"The personal future and peace corps competence." Journal of Personality and Social Psychology, Monograph Supplement, Vol. 8, No. 2, Part 2, pp. 1-26.

Falk, John, and Dalbir Bindra. 1954.
"Judgments of time as a function of serial position and stress." Journal of Experimental Psychology, 47:279-284.

Fang, Ge, Funi Fang, and Fan Liu. 1984.
"The development of children's cognition of the sequence of time: II." Acta Psychologica Sinica, 16(3):250-258.

Farber, Maurice L. 1944.
"Suffering and time-perspective of the prisoner." University of Iowa Studies in Child Welfare, 20:153-227.

Farber, Maurice L. 1953.
"Time-perspective and feeling tone: a study in the perception of the days." Journal of Psychology, 35:253-259.

Farmer, Eric W., Julie Hunter, and Andrew J. Belyavin. 1984.
"Performance under time constraint: the role of personality." Perceptual and Motor Skills, 59(3):875-884.

Farrell, Muriel. 1953.
"Understanding of time relations of five-, six- and seven-year-old children of high IQ." Journal of Educational Psychology, 46:587-594.

Faulkner, Kim K., and Stephen J. Duecker. 1989.
"Stress, time distortion, and failure to recover among obese individuals: implications for weight gain and dieting." International Journal of Eating Disorders, 8(2):247-250.

Feather, N.T., and M.J. Bond. 1983.
"Time structure and purposeful activity among employed and unemployed university graduates." Journal of Occupational Psychology, LVI, pp. 241-54.

Featherman, David L., and Trond Petersen. 1986.
"Markers of aging: modeling the clocks that time us." Research on Aging, 8(3):339-365.

Feifel, Herman. 1957.
"Judgment of time in younger and older persons." Journal of Gerontology, 12:71-74.

Felix, Morton. 1965.
"Time estimates as affected by incentive class and motivational level." Genetic Psychology Monographs, 72:353-399.

Fenchel, Gerd H. 1985.
"Time as self and ego experience." Issues in Ego Psychology, 8(1-2):73-81.

Ferguson, Robert P., and Martin Parthena. 1983.
"Long-term temporal estimation in humans." Perception and Psychophysics, 33(6):585-592.

Ferreiro, E., and H. Sinclair. 1971.
"Temporal relationships in language." International Journal of Psychology, 6:39-47.

Fetterman, J. Gregor, and Leon R. Dreyfus. 1986.
"Par comparison of durations." Behavioural Process, 12(2):111-123.

Fetterman, J. Gregor, D. Alan Stubbs, and Leon R. Dreyfus. 1986.
"Scaling of events spaced in time." Behavioural Process, 13(1-2):53-68.

Filer, Robert J., and Donald W. Meals. 1949.
"The effect of motivating conditions on the estimation of time." Journal of Experimental Psychology, 39:327-337.

Fink, Howard H. 1957.
"The relationship of time perspective to age, institutionalization, and activity." Journal of Gerontology, 12:414-417.

Fischhoff, Baruch, and Ruth Beyth. 1975.
"'I knew it would happen': remembered probabilities of once-future things." Organizational Behavior and Human Performance, 13(1):1-16.

Fisher, Seymour, and Rhoda Lee Fisher. 1953.
"Unconscious conception of parental figures as a factor influencing perception of time." Journal of Personality, 21:496-505.

Fivush, Robyn, and Jean M. Mandler. 1985.
"Developmental changes in the understanding of temporal sequence." Child Development, 56:1437-1446.

Flavell, John H. 1963.
The Developmental Psychology of Jean Piaget. New York: Van Nostrand.

Flexser, A.J., and G.H. Bower. 1974.
"How frequency affects recency judgments: a model for recency discrimination." Journal of Experimental Psychology, 86:137-143.

Fluckiger, Michelangelo. 1985.
"The effects of time pressure on duration

discrimination." In John A. Michon and Janet L. Jackson (eds.), Time, Mind, and Behavior, pp. 131-139. New York: Springer-Verlag.

Folkins, Carlyle H. 1970.
"Temporal factors and the cognitive mediators of stress reaction." Journal of Personality and Social Psychology, 14:173-184.

Fooken, Insa. 1982.
"Patterns of health behavior, life satisfaction, and future time perspective in a group of old aged women: data of 'survivors' from a longitudinal study on aging." International Journal of Behavioral Development, 5(3):367-390.

Fox, R.H., Pamela A. Bradsburry, I.G.H. Hampton, and C.F. Legg. 1967.
"Time judgment and body temperature." Journal of Experimental Psychology, 75(1):88-96.

Fraisse, Paul. 1963.
The Psychology of Time. New York: Harper and Row.

Fraisse, P. 1968.
"Time: psychological aspects." In David L. Sills (ed.), International Encyclopedia of the Social Sciences, Vol. 16, pp. 25-30. New York: MacMillan and Free Press.

Fraisse, Paul. 1978.
"Time and rhythm perception." In E.C. Carterette and M.P. Friedman (eds.), Handbook of Perception, Vol. 8, pp. 203-254. New York: Academic.

Fraisse, Paul. 1979.
"Time in psychology." In Frank Greenaway (ed.), Time and the Sciences, pp. 71-84. Paris, France: Unesco Press.

Fraisse, Paul. 1981.
"Cognition of time in human activity." In Gerry d'Ydewalle and Willy Lens (eds.), Cognition in Human Motivation and Learning, pp. 233-259. Hillsdale, NJ: Erlbaum.

Fraisse, Paul. 1982.
"The adaptation of the child to time." In William J.

Friedman (ed.), The Developmental Psychology of Time, pp. 113-140. New York: Academic.

Fraisse, Paul. 1984.
"Perception and estimation of time." Annual Review of Psychology, 35:1-36.

Frake, Charles O. 1985.
"Cognitive maps of time and tide among medieval seafarers." Man, 20(N.5.):254-270.

Frank, J., and J.E.R. Staddon. 1974.
"Effects of restraint on temporal discrimination behavior." Psychological Records, 24:123-130.

Frank, Lawrence K. 1939.
"Time perspectives." Journal of Social Philosophy, 4:293-312.

Frankenhauser, Marianne. 1959.
Estimation of Time: An Experimental Study. Stockholm, Sweden: Almqvist and Wiksell.

Frankenhauser, Marianne. 1960.
"Subjective time as affected by gravitational stress." Scandinavian Journal of Psychology, 1:1-6.

Fraser, J.T. 1989.
"The many dimensions of time and mind: an epistemic jigsaw puzzle game." In J.T. Fraser (ed.), Time and Mind: The Study of Time VI. Madison, CT: International Universities Press.

Frederickson, Charles G. 1988.
"Temporal experience: a two component model." Perceptual and Motor Skills, 66(1):63-68.

Fredericson, Emil. 1951.
"Time and aggression." Psychological Review, 58:41-51.

Freedman, Jonathan L., and Donald R. Edwards. 1988.
"Time pressure, task performance, and enjoyment." In Joseph E. McGrath (ed.), The Social Psychology of Time: New Perspectives, pp. 113-133. Newbury Park, CA: Sage.

Freire, Evelio, Bernard Gorman, and Alden E. Wessman. 1980.
"Temporal span, delay of gradification, and children's

socioeconomic status." _Journal of Genetic Psychology_, 137(2):247-55.

French, L.A., and A.L. Brown. 1977.
"Comprehension of _before_ and _after_ in logical and arbitrary sequences." _Journal of Child Language_, 4:247-256.

Frey, Dieter, Martin Kumpf, Martin Irle, and Gisla Gniech. 1984.
"Re-evaluation of decision alternatives dependent upon the reversibility of a decision and the passage of time." _European Journal of Social Psychology_, 14(4):447-450.

Fricke, Ruth. 1979.
"Orientation towards the future by Liberian schoolchildren: a contribution to the understanding of young West Africans." _Human Development_, 22(2):113-126.

Friedman, E.R. 1977.
"Judgments of time intervals by young children." _Perceptional and Motor Skills_, 45:715-720.

Friedman, William J. 1977.
"The development of children's understanding of cyclic aspects of time." _Child Development_, 48:1593-1599.

Friedman, William J. 1978.
"Development of time concepts in children." In H.W. Reese and L.P. Lipsitt (eds.), _Advances in Child Development and Behavior_, Vol. 12, pp. 167-298. New York: Academic.

Friedman, William J. (ed.). 1982.
The Developmental Psychology of Time. New York: Academic.

Friedman, William J. 1982.
"Conventional time concepts and children's structuring of time." In William J. Friedman (ed.), _The Developmental Psychology of Time_, pp. 171-208. New York: Academic.

Friedman, William J. 1983.
"Image and verbal processes in reasoning about the months of the year." _Journal of Experimental Psychology: Learning, Memory and Cognition_, 9(4):650-666.

Friedman, William J. 1984.
"Analog and semantic models of judgments about the months of the year." Memory and Cognition, 12(3):306-313.

Friedman, William J. 1986.
"The development of children's knowledge of temporal structure." Child Development, 57(6):1386-1400.

Friedman, William J., and P. Seely. 1976.
"The child's acquisition of spatial and temporal word meanings. Child Development, 47:1103-1108.

Friedman, William J., and Arnold J. Wilkins. 1985.
"Scale effects in memory for the time of events." Memory and Cognition, 13(2):168-175.

Friel, Charles M., and William T. Lhamon. 1965.
"Gestalt study of time estimation." Perceptual and Motor Skills, 21:603-606.

Fryer, D.M., and S. McKenna. 1986.
"The laying off of hands: unemployment and the experience of time." In S. Fineman (ed.), Unemployment: Personal and Social Consequences. London, UK: Tavistock.

Fuhrman, Robert W., and Robert S. Wyer. 1988.
"Event memory: temporal-order judgments of personal life experiences." Journal of Personality and Social Psychology, 54(3):365-384.

Gabrielsson, A. 1974.
"Performance of rhythm patterns." Scandinavian Journal of Psychology, 15:63-72.

Gagne, Eva E. 1979.
"Teacher's future time perspectives for students and curricula." Perceptual and Motor Skills, 48(1):138.

Galanter, Eugene. 1984.
"Timing of motor programs and temporal patterns: discussion paper." Annals of the New York Academy of Sciences, 423:193-197.

Galbraith, R.C. 1976.
"The effects of frequency and recency on judgments of frequency and recency." American Journal of Psychology, 89:515-526.

Galinat, Withold H., and Ingwer Borg. 1987.
"On symbolic temporal information: beliefs about the experience of duration." Memory and Cognition, 15(4):308-317.

Ganesan, Vedhagiri. 1982.
"Perceptions of self, village, and nation under three temporal dimensions." Psychological Research Journal, 6(1):7-15.

Gardner, Rick M., Beth Reyes, Shirley I. Brake, and Veronica E. Salaz. 1984.
"Reproduction and discrimination of time in obese subjects." Personality and Social Psychology, 10(4):564-573.

Garner, W.R., and R.L. Gottwald. 1967.
"Some perceptual factors in the learning of sequential pattern of binary events." Journal of Verbal Learning and Verbal Behavior, 6:582-589.

Garner, W.R., and R.L. Gottwald. 1968.
"The perception and learning of temporal patterns." Quarterly Journal of Experimental Psychology, 20:97-109.

Gee, J., and F. Grosjean. 1983.
"Performance structures: a psycholinguistic and linguistic appraisal." Cognitive Psychology, 15:411-458.

Gergen, Kenneth J., and Kurt W. Back. 1965.
"Aging, time perspective, and preferred solutions to international conflicts." Journal of Conflict Resolution, 9(2):177-186.

Getty, D.J. 1975.
"Discrimination of short temporal intervals: a comparison of two models." Perception and Psychophysics, 18:1-8.

Getty, D.J. 1976.
"Counting processes in human timing." Perception and Psychophysics, 20:191-197.

Gibbon, John. 1981.
"On the form and location of two psychometric bisection function of time." Journal of Mathematical Psychology, 24:58-87.

Gibbon, John. 1981.
"Two kinds of ambiguity in the study of psychological time." In M.L. Commons and J.A. Nevin (eds.), Quantitative Analysis of Behavior: Vol. 1. Discriminative Properties of Reinforcement Schedules, pp. 157-189. Cambridge, MA: Ballinger.

Gibbon, John, and Lorraine Allan (eds.). 1984.
Timing and Time Perception. New York: New York Academy of Sciences. (Vol. 423 of the Annals of the New York Academy of Sciences.)

Gibson, J.J. 1966.
"The problem of temporal order in stimulation and perception." Journal of Pyschology, 62:141-149.

Gibson, J.J. 1979.
The Ecological Approach to Visual Perception. Boston, MA: Houghton Mifflin.

Gilliland, A.R., J. Hofeld, and G. Eckstrand. 1946.
"Studies in time perception." Psychological Bulletin, 43:162-176.

Gjesme, Torgrim. 1975.
"Slope of gradients for performance as a function of achievement motive, goal distance in time, and future time orientation." Journal of Psychology, 91(1):143-160.

Gjesme, Torgrim. 1979.
"Future time orientation as a function of achievement motives, ability, delay of gratification, and sex." Journal of Psychology, 101(2):173-188.

Gjesme, Torgrim. 1981.
"Some factors influencing perceived goal distance in time: a preliminary check." Perceptual and Motor Skills, 53(1):175-182.

Gjesme, Torgrim. 1981.
"Is there any future in achievement motivation?" Motivation and Emotion, 5(2):115-138.

Gjesme, Torgrim. 1983.
"On the concept of future time orientation: considerations of some functions' and measurements'

implications." International Journal of Psychology, 18(5):443-461.

Glazer, H., and D. Singh. 1971. "Role of collateral behavior in temporal discrimination performance and learning." Journal of Experimental Psychology, 91:78-84.

Glenberg, A.M. 1976. "Monotonic and nonmonotonic lag effects in paired-associate and recognition memory paradigms." Journal of Verbal Learning and Verbal Behavior, 15:1-16.

Glenberg, A.M. 1979. "Component levels theory of the effects of spacing of repetitions on recall and recognition." Memory and Cognition, 7:95-112.

Goldfarb, Joyce L., and Sanford Goldstone. 1963. "Time judgment: a comparison of filled and unfilled durations." Perceptual and Motor Skills, 16:376.

Goldman, Amy P., and Frances Everett. 1985. "Delay of gratification and some time concept in reflective and impulsive children." Child Study Journal, 15(3):167-180.

Goldrich, Judith M. 1967. "A study in time orientation: the relation between memory for past experience and orientation to the future." Journal of Personality and Social Psychology, 6(2):216-221.

Goldstone, Sanford. 1967. "The human clock: a framework for the study of healthy and deviant time perception." Annals of the New York Academy of Sciences, 138:767-783.

Goldstone, Sanford, W.K. Boardman, and W.T. Lhamon. 1959. "Intersensory comparisons of temporal judgments." Journal of Experimental Psychology, 57:243-248.

Goldstone, Sanford, and Joyce L. Goldfarb. 1964. "Auditory and visual time judgments." Journal of General Psychology, 70:369-387.

Goldstone, Sanford, and Joyce L. Goldfarb. 1966. "The perception of time by children." In A.H. Kidd and J.L. Rivoire (eds.), _Perceptual Development in Children_, pp. 445-487. New York: International Universities Press.

Goldstone, Sanford, and William T. Lhamon. 1971. "Levels of cognitive functioning and the auditory visual differences in human timing behavior." In M.H. Appley (ed.), _Adaptation Level Theory: A Symposium_, pp. 263-280. New York: Academic.

Goldstone, Sanford, and William T. Lhamon. 1974. "Studies of the auditory-visual differences in human time judgments: I. Sounds are judged longer than lights." _Perceptual and Motor Skills_, 39:63-82.

Goldstone, Sanford, et al. 1963. "Sociometric status and apparent duration." _Journal of Social Psychology_, 61:303-310.

Gooddy, William. 1958. "Time and the nervous system: the brain as a clock." _Lancet_, 7031:1139-1141.

Gooddy, William. 1969. "Disorders of the time sense." In P.J. Vinken and G.W. Bruyn (eds.), _Handbook of Clinical Neurology: Disorders of Higher Nervous Activity_, Vol. 13, pp. 229-250. Amsterdam, Netherlands: North-Holland.

Goodman, Lisl Marburg. 1975. "Attitudes towards death in creative artists." _Omega: Journal of Death and Dying_, 6(4):345-356.

Gorman, Bernard S., and Alden E. Wessman (eds.). 1977. _The Personal Experience of Time_. New York: Plenum.

Gorman, Bernard S., and Alden E. Wessman (eds.). 1977. "Images, values, and concepts of time in psychological research." In Barnard S. Gorman and Alden E. Wessman (eds.), _The Personal Experience of Time_, 217-163. New York: Plenum.

Gorman, Bernard S., Alden E. Wessman, Gertrude R. Schmeidler, S. Thayer, and E. Mannucci. 1973. "Linear representation of temporal location and Stevens' law." _Memory and Cognition_, 1:169-171.

Green, D.M., and R. Duncan Luce. 1974.
"Variability of magnitude estimates: a timing theory analysis." Perception and Psychophysics, 15:291-300.

Green, Helen B. 1975.
"Temporal stages in the development of the self." In J.T. Fraser and N. Lawrence (eds.), The Study of Time II, pp. 1-19. New York: Springer-Verlag

Green, Helen B., and Robert H. Knapp. 1959.
"Time judgement, aesthetic preference, and need for achievement." Journal of Abnormal and Social Psychology, 58:140-142.

Greene, A.L. 1986.
"Future-time perspective in adolescence: The Present of Things Future revisited." Journal of Youth and Adolescence, 15(2):99-113.

Greeno, J., and H.A. Simon. 1974.
"Processes for sequence production." Psychological Review, 81:187-198.

Greer, K.L., and D.W. Green. 1983.
"Context and motor control in handwriting." Acta Psychologica, 54:205-215.

Grice, Stephen R. 1984.
"Time-order discrimination of sequences of four events." Annals of the New York Academy of Sciences, 423:592-593.

Grober, Ellen. 1984.
"Nonlinguistic memory in aphasia." Cortex, 20(1):67-73.

Guay, Michel, and C.R. Craig. 1984.
"Rehearsal of temporal visual information." Perceptual and Motor Skills, 58(1):215-220.

Guay, Michel, and Alan W. Salmoni. 1988.
"human time estimation: procedural effects." Bulletin of the psychonomic Society, 26(1):19-22.

Guenther, R.K., and M. Linton. 1975.
"Mechanisms of temporal coding." Journal of Experimental Psychology: Human Learning and Memory, 104:182-187.

Guilford, J.P. 1926.
"Spatial symbols in the apprehension of time." American Journal of Psychology, 37:420-423.

Guzy, L.T., and S. Axelrod. 1972.
"Interaural attention shifting as response." Journal of Experimental Psychology, 95:290-294.

Hamid, P., and L. James. 1973.
"Temporal perspectives and alienation." Perceptual and Motor Skills, 37:139-145.

Handel, Stephen. 1988.
"Space is to time as vision is to audition: seductive but misleading." Journal of Experimental Psychology: Human Perception and Performance, 14(2):315-317.

Hanstein, A.D., and G.S. Rotter. 1969.
"Research methodology in temporal perception." Journal of Experimental Psychology, 79:561-564.

Hare, R.D. 1963.
"Anxiety, temporal estimation, and rate of counting." Perception and Motor Skills, 16:441-444.

Harker, Judith O., and Walter H. Riege. 1985.
"Aging and delay of effects on recognition of words and designs." Journal of Gerontology, 40(5):601-604.

Harner, Lorraine. 1976.
"Yesterday and tomorrow: development of early understanding of the terms." Developmental Psychology, 11:864-865.

Harner, Lorraine. 1976.
"Young children's understanding of past and future reference." Journal of Psycholinguistic Research, 5:65-84.

Harner, Lorraine. 1980.
"Compensation of past and future reference revisited." Journal of Experimental Child Psychology, 29:170-182.

Harner, Lorraine. 1981.
"Children talk about the time and aspect of actions." Child Development, 52:498-506.

Harner, Lorraine. 1982.
"Talking about the past and the future." In William J. Friedman (ed.) The Developmental Psychology of Time, pp. 141-169. New York: Academic.

Harris, Paul, Gabriella R. Guz, Mark S. Upman, and Zhu Manshu. 1985.
"Insight into the time course of emotion among Western and Chinese children." Child Development, 56(4):972-988.

Harrison, M. 1934.
"The nature and development of concepts of time among young children." Elementary School Journal, 34:507-514.

Harton, John J. 1938.
"The influence of the difficulty of activity on the estimation of time." Journal of Experimental Psychology, 23:270-287.

Harton, John J. 1938.
"The influence of the difficulty of activity on the estimation of time. II." Journal of Experimental Psychology, 23:428-433.

Harton, John J. 1939.
"An investigation of the influence of success and failure on the estimation of time." Journal of General Psychology, 21:51-62.

Hasan, Abida. 1983.
"Punctuality and life satisfaction among working mothers as function of their age." Perspectives in Psychological Research, 6(1):42-43.

Hasher, L., and R.T. Zacks. 1979.
"Automatic and effortful processes in memory." Journal of Experimenal Psychology: General, 108:356-388.

Hawkins, Wendy L., Linda C. French, Bonnie D. Crawford, and Michael E. Enzle. 1988.
"Depressed affect and time perception." Journal of Abnormal Psychology, 97(3):275-280.

Hayes, Donald S., and B. Suzanne Kelly. 1984.
"Young children's processing of television: modality differences in the retention of temporal relations." Journal of Experimental Child Psychology, 38(3):505-514.

Healy, A.F. 1974.
"Separating item from order information in short-term memory." Journal of Verbal Learning and Verbal Behavior, 13:644-655.

Heath, Richard A. 1982.
"Memory for the order of sequential visual events." Australian Journal of Psychology, 34(3):267-279.

Heath, Richard A. 1984.
"Response time and temporal order judgment in vision." Australian Journal of Psychology, 36(1):21-34.

Heckel, Robert V., and Jayalakshmi Rajagopal. 1975.
"Future time perspective in Indian and American college students." Journal of Social Psychology, 95(1):131-132.

Heinemann, Eric G. 1984.
"A model for temporal generalization and discrimination." Annals of the New York Academy of Sciences, 423:361-371.

Hellstrom, Ake. 1977.
"Time errors are perceptual." Psychological Research, 39:345-388.

Hellstrom, Ake. 1985.
"The time-order error and its relatives: mirrors of cognitive processes in comparing." Psychological Bulletin, 97(1):35-61.

Helson, H., and S.M. King. 1931.
"The *tau* effect: an example of psychological reality." Journal of Experimental Psychology, 14:202-217.

Helson, R., et al. 1984.
"Personality and patterns of adherence and non-adherence to the social clock." Journal of Personality and Social Psychology, 46(5):1079-96.

Hendricks, J. (ed.). 1986.
"Aging and time." Special issue. American Behavioral Scientist, Vol. 29, No. 6.

Hendrikx, A. Paul. 1984.
"Temporal aspects of retrieval in short term serial retention." Acta Psychologica, 57(3):193-214.

Henik, Willy, and George Domino. 1975.
"Alterations in future time perspective in heroin addicts." Journal of Clinical Psychology, 31(3):557-564.

Henrickson, E.F. 1948.
"A study of stage-fright and the judgment of speaking time." Journal of Applied Psychology, 32:532-536.

Herman, James F., Lawra M. Norton, and Christine A. Klein. 1986.
"Children's distance estimates in a large-scale environment: a search for the route angularity effect." Environment and Behavior, 18(4):533-558.

Herman, James F., Lawra M. Norton, and Stephen F. Roth. 1983.
"Children's and adult's distance estimations in a large-scale environment: effects of time and clutter." Journal of Experimental Child Psychology, 36(3):453-470.

Herman, James F., Stephen F. Roth, and Lawra M. Norton. 1984.
"Time and distance in spatial cognition development." International Journal of Behavioral Development, 7(1):35-51.

Herriott, P. 1969.
"The comprehension of tense by young children." Child Development, 40:103-110.

Hicks, Robert E., Deborah A. Allen, and James P. Mayo. 1984.
"A developmental study of temporal duration judgments." Journal of Genetic Psychology, 144(1):31-38.

Hicks, Robert E., C. Thomas Gualtieri, James P. Mayo, and Mario Perez-Reyes. 1984.
"Cannabis, atropine, and temporal information processing." Neuropsychobiology, 12(4):229-237.

Hicks, Robert E., G.W. Miller, and M. Kinsbourne. 1976.
"Prospective and retrospective judgments of time as a function of amount of information processed." American Journal of Psychology, 89:719-730.

Hindle, H. 1951.
"Time estimates as a function of distance traveled and relative clarity of a goal." Journal of Personality, 19:483-501.

Hinrichs, J.V. 1970.
"A two-process memory-strength theory for judgment of recency." Psychological Review, 77:223-233.

Hinrichs, J.V., and H. Buschke. 1968.
"Judgment of recency under steady-state conditions." Journal of Experimental Psychology, 78:574-579.

Hintzman, D.L., J.J. Summers, and Richard A. Block. 1975.
"Spacing judgments as an index of study-phase retrieval." Journal of Experimental Psychology: Human Learning and Memory, 1:31-40.

Hirano, Shin. 1981.
"Effect of pause between discrete events on time estimation." Journal of Child Development, 17:35-39.

Hirsch, Bennett Z., and Janet M. Kulberg. 1987.
"Television and temporal development." Journal of Early Adolescence, 7(3):331-344.

Hirsh, I.J. 1974.
"Temporal order and auditory perception." In H.R. Moskowitz, B. Sharf, and J.C. Stevens (eds.), Sensation and Measurement, pp. 251-258. Dordrecht, Netherlands: Reidel.

Hirsh, I.J., and C.E. Sherrick, Jr. 1961.
"Perceived order in different sense modalities." Journal of Experimental Psychology, 62:423-432.

Hoagland, Hudson. 1933.
"The physiological control of the judgment of duration: evidence for chemical clock." Journal of General Psychology, 9:267-287.

Hockley, William E. 1984.
"Analysis of response time distributions in the study of cognitive processes." Journal of Experimental Psychology: Learning, Memory and Cognition, 10(4):598-615.

Hoffer, Abram. 1972.
"Problems connected with evaluation of the effects of psychedelic drugs on time perception." In Henri Yaker, Humphry Osmond, and Frances Cheek (eds.), The Future of Time: Man's Temporal Environment, pp. 389-404. Garden City, NY: Doubleday.

Hoffman, R.R., and J.M. Nead. 1983.
"General contextualism, ecological science and cognitive search." Journal of Mind and Behavior, 4:507-560.

Hogan, H. Wayne. 1978.
"A theoretical reconciliation of competing views of time perception." American Journal of Psychology, 91:417-428.

Holding, Dennis H., T.K. Noonan, H.D. Pfau, and Carol S. Holding. 1986.
"Date attribution, age, and the distribution of life-time memories." Journal of Gerontology, 41(4):481-485.

Hoopen, Gert Ten. 1985.
"The detection of anisochrony in monaural and interaural sequences." In John A. Michon and Janet L. Jackson (eds.), Time, Mind, and Behavior, pp. 140-150. New York: Springer-Verlag.

Hoopen, Gert Ten, and S. Akerboom. 1983.
"The subjective tempo difference between interaural and monaural sequences as a function of sequence length." Perception and Psychophysics, 34:465-469.

Hoopen, Gert Ten, and J. Vos. 1980.
"Attention switching is not a fatigable process: methodological comments on Axelrod and Guzy (1972)." Journal of Experimental Psychology: Human Perception and Performance, 6:180-183.

Hoopen, Gert Ten, and J. Vos. 1981.
"Attention switching and patterns of sound locations in counting clicks." Journal of Experimental Psychology: Human Perception and Performance, 7:342-355.

Hoopen, Gert Ten, J. Vos, and J. Dispa. 1982.
"Interaural and nonaural clicks and clocks: tempo difference versus attention switching." Journal of Experimental Psychology: Human Perception and Performance, 8:422-434.

Hoornaert, Johan. 1973.
"Time perspectives: theoretical and methodological considerations." Psychologica Belgica, 13(3):265-294.

Hopkins, Gordon W. 1984.
"Ultrastable stimulus-response latencies: towards a model of response-stimulus synchronization." Annals of the New York Academy of Sciences, 423:16-29.

Hopkins, Gordon W., and Alfred B. Kristofferson. 1980.
"Ultrastable stimulus-response latencies: acquisition and stimulus control." Perception and Psychophysics, 27:241-250.

Hopkins, Heather. 1986.
"Temporality and reflexivity: toward the creative engagement of consciousness." Human Relations, 39(7):635-645.

Hornstein, A.D., and G.S. Rotter. 1969.
"Research methodology in temporal perception." Journal of Experimental Psychology, 79:561-564.

Howard, J.H., and J.A. Ballas. 1982.
"Acquisition of acoustic pattern categories by exemplar observation." Organizational Behavior and Human Performance, 30:157-173.

Huang, I. 1943.
"Children's conception of physical causality: a critical summary." Journal of Genetic Psychology, 64:71-121.

Hughes, Elizabeth. 1979.
"Institutionalized older adults and their future orientation." Journal of the American Geriatrics Society, 27(3):130-134.

Hughes, R., and Mary E. Render. 1968.
"Estimates of psychological time among obese and non-obese women." Journal of Psychology, 32(6):654-658.

Hulbert, Ryan J., and Willy Lens. 1988.
"Time perspective, time attitude, and time orientation in alcoholism: a review." International Journal of the Addictions, 23(3):279-298.

Hulse, Stewart H., and N.P. Dorsky. 1977.
"Structural complexity as a determinant of serial pattern learning." Learning and Motivation, 8:488-506.

Hulstijn, W., and G.P. Van Galen. 1983.
"Programming in handwriting: reaction time and movement time as a function of sequence length." Acta Psychologica, 54:23-49.

Hultsch, David F., and Raymond W. Bortner. 1974.
"Personal time perspective in adulthood: a time-sequential study." Developmental Psychology, 10(6):835-837.

Humphrey, K., R.C. Tees, and J. Werker. 1979.
"Auditory-visual integration of temporal relation in infants." Canadian Journal of Psychology, 33:347-352.

Innis, N.K., and J.E.R. Staddon. 1971.
"Temporal tracking on cyclic-interval reinforcement schedules." Journal of the Experimental Analysis of Behavior, 16:411-423.

Israeli, Nathan. 1930.
"Some aspects of the social psychology of futurism." Journal of Abnormal and Social Psychology, 25:121-32.

Israeli, Nathan. 1932.
"The social psychology of time, comparative rating of emotional reactions to the past, present, and future." Journal of Abnormal and Social Psychology, 27:209-13.

Israeli, Nathan. 1933.
"Group predictions of future events." Journal of Social Psychology, 4:201-22.

Israeli, Nathan. 1934.
"Measurement of the outlook on the future." Psychological Bulletin, 31:643-644.

Israeli, Nathan. 1936.
Abnormal Personality and Time. New York: Science Press.

Jackson, Janet L. 1985.
"Is the processing of temporal information automatic or controlled?" In John A. Michon and Janet L. Jackson

(eds.), Time, Mind, and Behavior, pp. 179-190. New York: Springer-Verlag.

Jackson, Janet L. 1989.
"The processing og temporal information: do we indeed time our minds?" In J.T. Fraser (ed.), Time and Mind: The Study of Time VI. Madison, CT: International Universities Press.

Jackson, Janet L., and John A. Michon. 1984.
"Effects of item concreteness on temporal coding." Acta Psychologica, 57(2):83-95.

Jackson, Janet L., John A. Michon, and A. Vermeeren. 1984.
"The processing of temporal information." Annals of the New York Academy of Sciences, 423:603-604.

Jacoby, Jacob. 1968.
"Time perspective and dogmatism, a replication." Journal of Social Psychology, 79:281-282.

Jacquy, J., and P. Squelart. 1984.
"Right hemispheric activation on chronognosia: a preliminary study by rheoencephalography." Neuropsychobiology, 11(1):3-6.

Jahoda, G. 1963.
"Children's concepts of time and history." Educational Review, 15:87-104.

Jamieson, Donald G. 1977.
"Two presentation order effects." Canadian Journal of Psychology, 31:184-194.

Jamieson, Donald G, and William M. Petrusic. 1975.
"Presentation order effects during duration discrimination." Perception and Psychophysics, 17:197-202.

Jamieson, Donald G., and W. M. Petrusic. 1978.
"Feedback versus an illusion in time." Perception, 7:91-96.

Jamieson, Donald G., Elzbieta Slawinska, Margaret F. Cheeseman, and Blas Espinoza-Varas. 1984.
"Timing perturbations with complex auditory stimuli." Annals of the New York Academy of Sciences, 423:96-102.

Jasper, H., and C. Shagass. 1941. "Conscious time judgments related to conditioned time intervals and voluntary control of alpha rhythm." Journal of Experimental Psychology, 28:503-508.

Jenkins, J.J. 1974. "Remember that old theory of memory? Well, forget it!" American Psychologist, 29:785-795.

Jenkins, J.J. 1979. "Four points to remember: a tetrahedral model of memory experiments." In L.S. Cermak and F.I.M. Craik (eds.), Levels of Processing and Human Memory, pp. 429-446. Hillsdale, NJ: Erlbaum.

Jesteadt, W., R. Duncan Luce, and D.M. Green. 1977. "Sequential effects in judgments of loudness." Journal of Experimental Psychology: Human Perception and Performance, 3:92-104.

Johnson, J.E., and T.P. Petzel. 1971. "Temporal orientation and time estimation in chronic schizophrenics." Journal of Clinical Psychology, 27:194-196.

Johnston, Harriet M. 1960. "A comparison of time estimation of schizophrenic patients with that of normal individuals." Psychological Bulletin, 57:213-236.

Johri, G.N. 1984. "Subjective scaling in time estimation." Asian Journal of Psychology and Education, 13(2-4):37-40.

Jones, B. 1972. "Outflow and inflow in movement duplication." Perception and Psychophysics, 12:95-96.

Jones, Mari Riess. 1974. "Cognitive representations of serial patterns." In B. Kantowitz (ed.), Human Information Processing: Tutorials in Performance Cognition, pp. 187-229. Potomac, MD: Erlbaum.

Jones, Mari Riess. 1976. "Time, our lost dimension: toward a new theory of perception, attention, and memory." Psychological Review, 83:323-355.

Jones, Mari Riess. 1976.
"Levels of structure in the reconstruction of temporal and spatial serial patterns." Journal of Experimental Psychology: Human Learning and Memory, 2:475-488.

Jones, Mari Riess. 1981.
"A tutorial on some issues and method in serial pattern research." Perception and Psychophysics, 30:492-508.

Jones, Mari Riess. 1984.
"The patterning of time and its effects on perceiving." Annals of the New York Academy of Sciences, 423:158-167.

Jones, Mari Riess. 1985.
"Structural organization of events in time." In John A. Michon and Janet L. Jackson (eds.), Time, Mind, and Behavior, pp. 192-214. New York: Springer-Verlag.

Jones, Mari Riess, Marilyn Boltz, and G. Kidd. 1982.
"Controlled attending as a function of melodic and temporal context." Perception and Psychophysics, 32:211-218.

Jones, Mari Riess, G. Kidd, and R. Wetzel. 1981.
"Evidence for rhythmic attention." Journal of Experimental Psychology: Human Perception and Performance, 7:1057-1073.

Jones, Mari Riess, D.J. Maser, and G.R. Kidd. 1978.
"Rate and structure in memory for auditory patterns." Memory and Cognition, 6:246-258.

Jones, Mari Riess, and K.P. Zamostny. 1975.
"Memory and rule structure in the prediction of serial patterns." Journal of Experimental Psychology: Human Learning and Memory, 1:295-306.

Joubert, Charles E. 1983.
"Subjective acceleration of time: death anxiety and sex differences." Perceptual and Motor Skills, 57(1):49-50.

Joubert, Charles E. 1984.
"Structured time and subjective acceleration of time." Perceptual and Motor Skills, 59(1):335-336.

Judson, Abe J., and C.E. Tuttle. 1966.
"Time perspective and social class." Perceptual and Motor Skills, 23:1074.

Kane, Linda and Bradley A. Lown. 1986.
"Steven's power law and time perception: effect of filled intervals, duration of the standard and number of presentations of the standard." Perceptual and Motor Skills, 62(1):35-38.

Kastenbaum, Robert J. 1961.
"The dimensions of future time perspective: an experimental analysis." Journal of General Psychology, 65:203-218.

Kastenbaum, Robert J. 1963.
"Cognitive and personal futurity in later life." Journal of Individual Psychology, 19:216-222.

Kastenbaum, Robert J. 1964.
"The structure and function of time perspectives." Journal of Psychological Researches, 8:97-105.

Kastenbaum, Robert J. 1965.
"The direction of time perspective: I. The influence of affective set." Journal of General Psychology, 73:189-201.

Kastenbaum, Robert J. 1965.
"Engrossment and perspective in later life: a developmental field approach." In Robert J. Kastenbaum (ed.), Contributions to the Psychobiology of Aging, pp. 3-18. New York: Springer.

Kastenbaum, Robert J. 1966.
"On the meaning of time in later life." Journal of Genetic Psychology, 109:9-25.

Kastenbaum, Robert J. 1967.
"The impact of experience with the aged upon the time perspective of young adults." Journal of Genetic Psychology, 110(2):159-167.

Kastenbaum, Robert J. 1969.
"The foreshortened life perspective." Geriatrics, 24:126-133.

Kastenbaum, Robert J. 1975.
"Time, death and ritual in old age." In J.T. Fraser and N. Lawrence (eds.), The Study of Time II, pp. 20-38. New York: Springer-Verlag.

Kastenbaum, Robert J. 1977.
"Memories of tomorrow: on the interpenetrations of time in later life." In Bernard S. Gorman and Alden E. Wessman (eds.), The Personal Experience of Time, pp. 193-214. New York: Plenum.

Kausler, Donald H., Wemara Lichty, and Roger T. Davis. 1985.
"Temporal memory for performed acitivites: intentionality and adult age differences." Developmental Psychology, 21(6):1132-1138.

Keele, Steven W., Robert A. Pokorny, Daniel M. Corcos, and Richard Ivry. 1985.
"Do perception and motor production share common timing mechanisms: a correlational analysis." Acta Psychologica, 60(2-3):173-191.

Keen, Ernest. 1978.
"The past in the future: consciousness and tradition." Journal of Humanistic Psychology, 18(4):5-18.

Keith, P.M. 1981/82.
"Perceptions of time remaining and distance from death." Omega, 12(4):307-318.

Kelly, G.A. 1955.
The Psychology of Personal Constructs, Vols. I and II. New York: W.W. Norton.

Kelly, H. 1973.
"The process of causal attribution." American Psychologist, 28:107-128.

Kelly, Janice R. 1988.
"Entrainment in individual and group behavior." In Joseph E. McGrath (ed.), The Social Psychology of Time: New Perspectives, pp. 89-110. Newbury Park, CA: Sage.

Kelly, Janice R., and Joseph E. McGrath. 1985.
"Effects of time limits and task types on task performance and interaction of four-person groups." Journal of Personality and Social Psychology, 49(2):395-407.

Kendall, Martha, and R.F. Sibley. 1970.
"Social class differences in time orientation: artifact?" Journal of Social Psychology, 82:187-91.

Kerr, W.A., and R.C. Keil. 1963.
"A theory and factory experiment on the time-drag concept of boredom." Journal of Applied Psychology, 47:7-9.

Ketchum, J.D. 1951.
"Time, values, and social organization." Canadian Journal of Psychology, 5:97-109.

Khoury, Robert M., and Gerald T. Thurmond. 1978.
"Ethnic differences in time perception: a comparison of Anglo and Mexican Americans." Perceptual and Motor Skills, Vol. 47(3, Pt. 2):1183-1188.

Kiernat, Jean M. 1983.
"Retrospection as a life span concept." Physical and Occupational Therapy in Geriatrics, 3(2):35-48.

Kikkawa, Masao. 1983.
"The effect of distribution of attention upon prospective and retrospective estimations of long temporal intervals." Journal of Child Development, 19:14-22.

Killeem, Peter R. 1975.
"On the temporal control of behavior." Psychological Review, 82:89-115.

Killeen, Peter R. 1984.
"Incentive theory III: adaptive clocks." Annals of the New York Academy of Sciences, 423:515-527..

Kimberly, R.P. 1970.
"Rhythmic patterns in human interaction." Nature, 228:88-90.

Kirkcaldy, Bruce D. 1984.
"Individual differences in time estimation." International Journal of Sport Psychology, 15(1):11-24.

Kitamura, Toshinori, and Rajinder Kumar. 1984. "Controlled study on time reproduction of depressive patients." Psychopathology, 17(1):24-27.

Klapp, S.T. 1979. "Doing two things at once: the role of temporal compatibility." Memory and Cognition, 7:375-381.

Kleber, R.J., William T. Lhamon, and Sanford Goldstone. 1963. "Hyperthermia, hyperthyroidism, and time judgment." Journal of Comparative Physiological Psychology, 56:362-365.

Klein, K.E., H. Bruner, et al. 1972. "Psychological and physiological changes caused by desynchronization following transzonal air travel." In W.P. Colquhoun (ed.), Aspects of Human Efficiency, pp. 295-305. London, UK: English Universities Press.

Klein, R., and R. Armitage. 1979. "Rhythms in human performance: 1 1/2 hour oscillations in cognitive style." Science, 204:1326-1328.

Klineberg, Stephen. L. 1967. "Changes in the outlook on the future between childhood and adolscence." Journal of Personality and Social Psychology, 7:185-193.

Klineberg, Stephen L. 1968. "Future time perspective and the preference for delayed reward." Journal of Personality and Social Psychology, 8(3,Pt.1):253-257.

Knapp, Robert H. 1972. "Personality and the psychology of time." In J.T. Fraser, F.C. Haber, and G.H. Muller (eds.), The Study of Time, pp. 312-319. New York: Springer-Verlag.

Knapp, Robert H., and John T. Garbutt. 1958. "Time imagery and achievement motive." Journal of Personality, 26:426-434.

Knapp, Robert H., and Helen B. Green. 1961. "The judgment of music filled intervals and need achievement." Journal of Social Psychology, 54:263-267.

Koenig, Fredrick. 1979.
"Future orientation and external locus on control." Psychological Reports, 44(3):957-958.

Koenig, Fredrick, William Swanson, and Carl Harter. 1980.
"Future orientation and social status." Perceptual and Motor Skills, 51(3, Pt. 1):927-930.

Koenig, Fredrick, William Swanson, and Carl Harter. 1981.
"Future time orientation, social class and anomia." Social Behavior and Personality, 9(2):123-127.

Koestler, Arthur. 1972.
The Roots of Coincidence. New York: Random House.

Kok, Gerjo. 1983.
"The further away, the less serious: effects of temporal distance on perceived value and probability of a future event." Psychological Reports, 52(2):531-535.

Koopmans, T.C. 1972.
"Representations of preference orderings over time." In C.B. McGuire and R. Radner (eds.), Decisions and Organizations, pp. 79-100. Amsterdam, Netherlands: North Holland.

Koothur, Joicy P., and S. Kanekar. 1984.
"Type A behaviour pattern and academic performance of college students in India." Journal of Social Psychology, 123(2):275-276.

Koriat, A., and Baruch Fischhoff. 1974.
"What day is today? An inquiry into the process of time orientation." Memory and Cognition, 2:201-205.

Kornfeld, Alfred D., and Patricia E. Marshall. 1987.
"SAT and TAT scores as measures of time perspective in institutionalized and community-based senior adults." International Journal of Psychosomatics, 34(3):11-13.

Kotovsky, K., and H.A. Simon. 1973.
"Empirical tests of a theory of human acquisition of concepts for sequential patterns." Cognitive Psychology, 4:399-424.

Kowal, Kathleen. 1981.
"Growth apparent duration: effect of melodic and non-

melodic tonal variation." Perceptual and Motor Skills, 52:803-817.

Kowal, Kathleen. 1984.
"Familiar melodies seem shorter, not longer, when played backwards: data and theory." Annals of the New York Academy of Sciences, 423:610-611.

Kowal, S., D.C. O'Connell, E.A. O'Brien, and E.T. Bryant. 1975.
"Temporal aspects of reading aloud and speaking: three experiments." American Journal of Psychology, 88:549-569.

Krauss, Herbert H. 1967.
"Anxiety: the dread of a future event." Journal of Individual Psychology, 23:88-93.

Krishnan, Lila, and Neenlaksh K. Saxena. 1984.
"Perceived time: its relationship with locus of control, filled versus unfilled time intervals, and perceiver's sex." Journal of General Psychology, 110(2):275-281.

Kristofferson, Alfred B. 1967.
"Attention and psychophysical time." Acta Psychologica, 27:93-100.

Kristofferson, Alfred B. 1976.
"Low-variance stimulus-response latencies: deterministic interval delays?" Perception and Psychophysics, 20:89-100.

Kristofferson, Alfred B. 1977.
"A real-time criterion theory of duration discrimination." Perception and Psychophysics, 21:105-117.

Kristofferson, Alfred B. 1980.
"A quantal step function in duration discrimination." Perception and Psychophysics, 27:300-306.

Kristofferson, Alfred B. 1984.
"Quantal and deterministic timing in human duration determination." Annals of the New York Academy of Sciences, 423:3-15.

Krus, David, and Stephen H. Fletcher. 1986. "Time: a speeding train or wind-driven sand? The estimation of fixed temporal intervals as related to images of time." Perceptual and Motor Skills, 62(3):936-938.

Kuch, D.O. 1974. "Differentiation of press durations with upper and lower limits on reinforced values." Journal of the Experimental Analysis of Behavior, 22:275-283.

Kuczaj, S. 1977. "The acquisition of regular and irregular past tense forms." Journal of Verbal Learning and Verbal Behavior, 16:589-600.

Kuhn, D., and H. Phelps. 1976. "The development of children's comprehension of causal direction." Child Development, 47:248-251.

Kun, A. 1978. "Evidence for preschooler's understanding of causal direction in extended causal sequences." Child Development, 49:218-222.

Lacquaniti, F., C. Terzuolo, and P. Viviani. 1983. "The law relating to the kinematic and figural aspects of drawing movements." Acta Psychologica, 54:115-130.

Lamm, Helmut, Rolf W. Schmidt, and Gisela Trommsdorff. 1976. "Sex and social class as determinant of future orientation (time perspective) in adolescents." Journal of Personality and Social Psychology, 34(3):317-326.

Landau, Simha F. 1975. "Future time perspective of delinquents an non-delinquents: the effect of institutionalization." Criminal Justice and Behavior, 2(1):22-36.

Langer, Jonas, Seymour Wapner, and Heinz Werner. 1961. "The effect of danger upon the experience of time." American Journal of Psychology, 74:94-97.

Langley, David J., Howard N. Zelaznik. 1984. "The acquisition of time properties associated with a sequential motor skill." Journal of Motor Behavior, 16(3):275-301.

La Roche, A.N., and A. Frankel. 1986. "Time perspective and health." Health Education Research, 1(2):139-142.

Lassen, G.L., T.C. Daniel, and N.R. Bartlett. 1974. "Judgment of recency for pictures and words." Journal of Experimental Psychology, 102:795-798.

Latzer, Robert W. 1973. "Nondirected light signals and the structure of time." Synthese, 24:236-280.

Lawton, M. Powell, Miriam Moss, and Mark Fulcomer. 1986. "Objective and subjective uses of time by older people." International Journal of Aging and Human Development, 24(3):171-188.

Lea, S.E.G., and S.M. Dow. 1984. "The integration of reinforcements over time." Annals of the New York Academy of Sciences, 423:269-277.

Lee, C., and J.M. Innes. 1983. "Type A behaviour pattern, time urgency and arrival time: a replication." Perceptual and Motor Skills, 56:177-178.

Lee, C.L., and William K. Estes. 1977. "Order and position in primary memory for letter strings." Journal of Verbal Learning and Verbal Behavior, 16:395-418.

Lee, C.L., and William K. Estes. 1981. "Item and order information in short-term memory: evidence for multi-level perturbation processes." Journal of Experimental Psychology: Human Learning and Memory, 7:149-169.

Lee, D., J. Lishman, and J. Thomson. 1982. "Regulation of gait in long jumping." Journal of Experimental Psychology: Human Perception and Performance, 8:448-459.

Lee, D.N., et al. 1983. "Visual timing in hitting an accelerating ball."

Quarterly Journal of Experimental Psychology: Human Experimental Psychology, 35A(2):333-346.

Lee, Seong-Soo. 1985.
"Children's acquisition of conditional logic structure: teachable?" Contemporary Education Psychology, 10(1):14-27.

Leeuwenberg, E.L.J. 1969.
"Quantitative specification of information in sequential patterns." Psychological Review, 76:216-220.

Legg, C.F. 1968.
"Alpha rhythm and time judgment." Journal of Experimental Psychology, 78(1):46-49.

Lehmann, Heinz. 1967.
"Time and psychopathology." Annals of the New York Academy of Sciences, 138:798-821.

Lehr, Ursula. 1967.
"Attitudes towards the future in old age." Human Development, 10(3-4):230-238.

Lens, Willy. 1975.
"Sex differences in attitude towards personal past, present, and future." Psychologica Belgica, 15(1):29-33.

Lens, Willy, and Antoine Gailly. 1980.
"Extension of future time perspective in motivational goals of different age groups." International Journal of Behavioral Development, 3(1):1-17.

Le Shan, Lawrence L. 1952.
"Time orientation and social class." Journal of Abnormal and Social Psychology, 47:589-592.

Lessing, Elise E. 1968.
"Demographic, developmental, and personality correlates of length of future time perspective." Journal of Personality, 36(2):183-201.

Lessing, Elise E. 1971.
"Comparative extension of personal and social-political future time perspective." Perceptual and Motor Skills, 33(2):415-422.

Lessing, Elise E. 1972.
"Extension of personal future time perspective, age, and life satisfaction of children and adolescents." Developmental Psychology, 6(3):457-468.

Levin, Iris. 1977.
"The development of time concepts in young children: reasoning about duration." Child Development, 48:435-444.

Levin, Iris. 1979.
"Interference of time-related and unrelated cues with duration comparisons of young children: analysis of Piaget's formulation of the relation of time and speed." Child Development, 50:469-477.

Levin, Iris. 1982.
"The nature and development of time concepts in children: the effects of interfering cues." In William J. Friedman (ed.), The Developmental Psychology of Time, pp. 47-85. New York: Academic.

Levin, Iris, I. Gilat, and T. Zelniker. 1980.
"The role of cue salience in the development of time concepts: duration comparisons in young children." Developmental Psychology, 16:661-671.

Levin Iris, and Tamar Globerson. 1984.
"The development of time concepts among advantaged and disadvantaged Israeli children." Journal of Genetic Psychology, 145(1):117-125.

Levin, Iris, Roberta Goldstein, and T. Zelniker. 1984.
"The role of memory and integeration in early time concepts." Journal of Experimental Child Psychology, 37(2):262-270.

Levin, Iris, E. Israeli, and E. Darom. 1978.
"The development of time concepts in young children: the relation between duration and succession." Child Development, 49:755-764.

Levin, Iris, and Helen Simons. 1986.
"The nature of children's and adult's concepts of time, speed, and distance and their sequence in development: analysis via circular motion." In Iris Levin (ed.),

Stage and Structure: Reopening the Debate, pp. 77-105. Norwood, NJ: Ablex.

Levin, Iris, Friedrich Wilkening, and Dembo Yoram. 1984. "Development of time quantification: integration and nonintegration of beginnings and endings in comparing durations." Child Development, 55(6):2160-2172.

Levine, Robert V., and K. Bartlett. 1984. "Pace of life, punctuality, and coronary heart disease in six countries." Journal of Cross-Cultural Psychology, 15:233-55.

Levine, Robert V., L.J. West, and H.T. Reis. 1980. "Perceptions of time and punctuality in the United States and Brazil." Journal of Personality and Social Psychology, 38:541-550.

Levitt, Eugene E. 1983. "Estimating the duration of sexual behaviour: a laboratory analog study." Archives of Sexual Behaviour, 12(4):329-335.

Levy, Sandra M. 1978-79. "Temporal experience in the aged: body integrity and the social milieu." International Journal of Aging and Human Development, 9(4):313-344.

Lhamon, William T., and Sanford Goldstone. 1960. "The time sense." Psychological Bulletin, 57:213-236.

Lhamon, William T., Sanford Goldstone, and Joyce L. Goldfarb. 1960. "The psychopathology of time judgment." In P. Hoch and J. Zubin (eds.), The Psychopathology of Perception, pp. 164-188. New York: Grune and Stratton.

Lichenstein, M. 1961. "Phenomenal simultaneity with irregular timing of components of the visual stimulus." Perceptual and Motor Skills, 12:47-60.

Lichtenberg, Philip. 1956. "Time perspective and the initiation of cooperation." Journal of Social Psychology, 43:247-60.

Linton, M. 1978.
"Real world memory after six years: an in vivo study of very long term memory." In M.M. Gruneberg, P.E. Morris, and R.N. Sykes (eds.), Practical Aspects of Memory, pp. 69-76. London, UK, Academic.

Litchfield, Peter M., and Jerome M. Sattler. 1968.
"An hypothesis: the existential notion of intentional time as a dimension of psychological health." Journal of Genetic Psychology, 79:257-270.

Lockhart, J.M. 1967.
"Ambient temperature and time estimation." Journal of Experimental Psychology, 73(2):286-291.

Lockhart, R.S. 1969.
"Recency discrimination predicted from absolute lag judgments." Perception and Psychophysics, 6:42-44.

Loehlin, J.C. 1959.
"The influence of different activities on the apparent length of time." Psychological Monographs, 474, No. 4, pp. 70-76.

Loehlin, J.C. 1960.
"The influence of different activities on the perception of time." Psychological Bulletin, 57:213-236.

Loftus, Elizabeth F., and Wesley Marburger. 1983.
"Since the eruption of Mt. St. Helens, has anyone beaten you up? Improving the accuracy of retrospective reports with landmark events." Memory and Condition, 11(2):114-120.

Lomranz, Jacob. 1983.
"Time estimation as a function of stimulus complexity and personality." Social Behavior and Personality, 11(2):77-82.

Lomranz, Jacob, Ariella Friedman, George Gitter, Dov Shmotkin, and Gedon Medini. 1985.
"The meaning of time related concepts across the life span: an Israeli sample." International Journal of Aging and Human Development, 21(2):87-107.

Lomranz, Jacob, Gedon Medini, and Ron Aschuach. 1982. "Realism as a cognitive indicator of self-actualization in temporality." Journal of Psychology, 110(1):53-62.

Lomranz, Jacob, Dov Shmotkin, and D.B. Katznelson. 1983. "Coherence as a measure of future time perspective in children and its relationship to delay of gratification and social class." International Journal of Psychology, 18(5):407-413.

Longobardi, E.T., and P.A. Wolff. 1973. "A comparison of motoric and verbal responses on a Piagetian rate-time task." Child Development, 44:433-437.

Lopata, Helena Z. 1986. "Time in anticipated future and events in memory." American Behaviorial Scientist, 29(6):695-709.

Low, Jaclyn F. 1987. "Time perception and rehabilitation of the elderly." Physical and Occupational Therapy in Geriatrics, 5(4):17-30.

Lowe, F., and P. Harzem. 1977. "Species differences in temporal regulation of behavior." Journal of the Experimental Analysis of Behavior, 28:189-201.

Luce, R. Duncan. 1984. "Time perception: discussion paper." Annals of the New York Academy of Sciences, 423:78-81.

Lundberg, Ulf, and Erkki Ellonen. 1977. "Involvement in potential future events estimated by males and females." Scandinavian Journal of Psychology, 18(4):333-338.

Lundberg, Ulf, Johan M. Von Wright, Marianne Frankenhaeuser, and Ulf-Johan Olsen. 1974. "Notes on involvement in future events as a function of temporal distance." Perceptual and Motor Skills, 39(2):841-842.

Lundberg, Ulf, Johan M. Von Wright, Marianne Frankenhaeuser, and Ulf-Johan Olsen. 1975.

"Involvement in four future events as a function of temporal distance." Scandinavian Journal of Psychology, 16(1):2-6.

Lynch, Denis J. 1971.
"Future time perspective and impulsivity in old age." Journal of Genetic Psychology, 118(2):245-252.

Lynch, William F. 1985.
"Imagining past, present, future in one piece." Studies in Formative Spirituality, 6(1):65-72.

Lynn, R. 1961.
"Introversion-extraversion differences in judgements of time." Journal of Abnormal and Social Psychology, 63(2):457-458.

Macar, Francoise. 1985.
"Time psychophysics and related models." In John A. Michon and Janet L. Jackson (eds.), Time, Mind, and Behavior, pp. 112-130. New York: Springer-Verlag.

Macar, Francoise, and U. Besson. 1985.
"Contingent negative variation in process of expectancy, motor preparation and time estimation." Biological Psychology, 21(4):293-307.

MacKenzie, Andrew. 1975.
Riddle of the Future: A Modern Study of Precognition. New York: Taplinger.

MacLeod, R.B., and M.T. Roff. 1936.
"An experiment in temporal disorientation." Acta Psychologica, 1:381-423.

Malhotra, Rekha, and K.D. Broota. 1984.
"Perception of brief durations as a function of size." Psychological Studies, 29(2):125-130.

Mandes, Evans. 1984.
"Effects of laterality and visual angle on time judgments: a phenomenological investigation of time durations." Perceptual and Motor Skills, 58(1):307-310.

Mandler, J. 1979.
"Categorical and schematic organization in memory." In

C.R. Puff (ed.), _Memory Organization and Structure_, pp. 259-289. New York: Academic.

Manger, Terje, and Karl H. Teigen. 1988. "Time horizon in students' predictions of grades." _Scandinavian Journal of Educational Research_, 32(2):77-91.

Mann, Harriet, Miriam Siegler, and Humphry Osmond. 1968. "The many worlds of time." _Journal of Analytical Psychology_, 13(1):33-56.

Mann, Harriet, Miriam Siegler, and Humphry Osmond. 1972. "The psychotypology of time." In Henri Yaker, Humphry Osmond, and Frances Cheek (eds.), _The Future of Time: Man's Temporal Environment_, pp. 142-178. Garden City, NY: Doubleday.

Marquis, D.P. 1941. "Learning in the neonate." _Journal of Experimental Psychology_, 29:263-282.

Martin, Donald S., and Ming-Shiunn Huang. 1984. "Effects of time and perceptual orientation on actors' and observers' attributions." _Perceptual and Motor Skills_, 58(1):23-30.

Mass, Jeannette P. 1985. "Beliefs relating to temporal dimensions held by three ethnic groups in the Fiji Islands." _Psychological Studies_, 30(1):53-65.

Massarik, Fred. 1968. "'Today' as an integrating factor." In Charlotte Buhler and Fred Massarik (eds.), _The Course of Human Life_, pp. 393-399. New York: Springer.

Massaro, Dominic W. 1975. _Experimental Psychology and Information Processing_. Chicago, IL: David MacNally.

Massaro, Dominic W. 1976. "Perceiving and counting sounds." _Journal of Experimental Psychology: Human Perception and Performance_, 2:337-346.

Massaro, Dominic W. 1984.
"Time's role for information, processing and normalization." _Annals of the New York Academy of Sciences_, 423:372-384.

Massaro, Dominic W., and W.L. Idson. 1976.
"Temporal course of perceived auditory duration." _Perception and Psychophysics_, 20:331-352.

Massaro, Dominic W., and W.L. Idson. 1978.
"Target-mask similarity in backward recognition masking of perceived tone duration." _Perception and Psychophysics_, 24:235-236.

Matsuda, Fumiko. 1966.
"Development of time estimation: effects of frequency of sounds given during standard time." _Japanese Journal of Psychology_, 36:285-294.

Matsuda, Fumiko. 1977.
"Effects of time and velocity on path-length estimation in children and adults." _Studia Psychologica_, 19:19-31.

Matsuda, Fumiko, and Michihiko Matsuda. 1977.
"Effects of spatial separation as a cue of time estimation in children and adults." _Japanese Psychological Research_, 21:132-138.

Matsuda, Fumiko, and Michihiko Matsuda. 1983.
"A longitudinal study of learning process of duration estimation in young children." _Japanese Psychological Research_, 25(3):119-129.

McCarthy, S. Viterbo. 1986.
"Perception of life after seventy." _Activities, Adaptation and Aging_, 8(2):39-47.

McClain, Lucida. 1983.
"Effect of processing demands on prospective and retrospective reports." _Perception and Psychophysics_, 34(2):185-189.

McClelland, J.L. 1979.
"On the time relations of mental processes: an examination of processes in cascade." _Psychological Review_, 86:287-330.

McConchie, R.D., and J. Rutschmann. 1971.
"Human time estimation: on differences between methods." _Perceptual and Motor Skills_, 32:319-336.

McConckie, George W., N. Roderick Underwood, D. Zola, and G.S. Wolverton. 1985.
"Some temporal characteristics of processing during reading." _Journal of Experimental Psychology: Human Perception and Performance_, 11(2):168-186.

McDermott, D. 1982.
"A temporal logic for reasoning about processes and plans." _Cognitive Science_, 6:101-155.

McGrath, Joseph E. 1986.
"Continuity and change: time, method, and the study of social issues." _Journal of Social Issues_, 42(2):5-19.

McGrath, Joseph E. 1988.
Research in the Social Psychology of Time. Newbury Park, CA: Sage.

McGrath, Joseph E. (ed.). 1988.
The Social Psychology of Time: New Perspectives. Newbury Park, CA: Sage.

McGrath, Joseph E., and Janice R. Kelly. 1986.
Time and Human Interaction: Toward a Social Psychology of Time. New York: Guilford Press.

McGrath, Joseph E., Janice R. Kelly, and Diane E. Machatka. 1984.
"The social psychology of time: entrainment of behavior in social and organizational settings." _Applied Social Pyschology Annual_, 5:21-44.

McGraw, Kathleen M., and Tom R. Tyler. 1986.
"The threat of nuclear war and psychological well-being." _International Journal of Mental Health_, 15(1-3):172-188.

McGregor, Douglas. 1938.
"The major determinants of the prediction of social events." _Journal of Abnormal and Social Psychology_, 38:179-204.

McLeod, Ronald W., H.E. Ross. 1983.
"Optic flow and cognitive factors in time-to-collision estimates." Perception, 12(4):417-423.

McNally, K.A., and Stephen Handel. 1977.
"Effect of element composition on streaming and the ordering or repeating sequences." Journal of Experimental Psychology: Human Perception and Performance, 3:451-460.

McNutt, T.H., and K.B. Melvin. 1968.
"Time estimation in normal and retarded subjects." American Journal of Mental Deficiency, 72:582-589.

McReynolds, P. 1953.
"Thinking conceptualized in terms of interacting moments." Psychological Review, 60:319-330.

Meade, Robert D. 1960.
"Time estimates as affected by need tension, and rate of progress." Journal of Psychology, 50:173-177.

Meade, Robert D. 1963.
"Effect of motivation and progress on the estimation of long time intervals." Journal of Experimental Psychology, 65(6):564-567.

Meade, Robert D. 1972.
"Future time perspective of Americans and subcultures of India." Journal of Cross-Cultural Psychology, 3(1):93-99.

Meade, Robert D., and L. Singh. 1968.
"Psychological time in India and America." Journal of Social Psychology, 76(2):169-74.

Meade, Robert D., and L. Singh. 1970.
"Motivation and progress effects on psychological time in subcultures of India." Journal of Social Psychology, 80(1):3-10.

Meade, Robert D., and L. Singh. 1971.
"Future time perspectives of college students in America and in India." Journal of Social Psychology, 83(2):175-182.

Meck, Warren H., and Russell M. Church. 1982.
"Discrimination of intertrial intervals in cross-modal transfer of duration." Bulletin of the Psychonomic Society, 19:234-236.

Meck, Warren H., Russell M. Church, and David S. Olton. 1984.
"Hippocampus, time, and memory." Behavioral Neuroscience, 98(1):3-22.

Mehta, Perin H., Pritam K. Rohila, Norman D. Sundberg, and Leona E. Tyler. 1972.
"Future time perspectives of adolescents in India and the United States." Journal of Cross-Cultural Psychology, 3(3):293-302.

Meisenhelder, Thomas M. 1985.
"An essay on time and the phenomenology of imprisonment." Deviant Behavior, 6(1):39-56.

Melikan, Levon H. 1969.
"Acculturation, time perspective, and feeling tone: a cross-cultural study in the perception of the days." Journal of Social Psychology, 79:273-275.

Melikan, Levon H. 1959.
"Preference for delayed reinforcement: an experimental study among Palestine Arab refugee children." Journal of Social Psychology, 79, pp. 273-275.

Melka, Raymond. 1969.
"Punctuality: an inquiry into the psychology of modern man." Diogenes, 65:20-48.

Mendelson, R., and T. Shultz. 1976.
"Covariation and temporal contiguity as principles of causal inference in young children." Journal of Experimental Child Psychology, 22:408-412.

Mentzer, Carole A., and Julie A. Schorr. 1986.
"Perceived situational control and perceived duration of time: expressions of life patterns." Advances in Nursing Science, 9(1):12-20.

Messe, Lawrence A., Roger W. Buldain, and Barbara Watts. 1981.

"Recall of social events with the passage of time." Personality and Social Psychology Bulletin, 7(1):33-38.

Meyer, David E., Steven Yantis, Allen M. Osman, and J. Keith Smith. 1985.
"Temporal properties of human information processing: tests of discrete versus continuous models." Cognitive Psychology, 17(4):445-518.

Michon, John A. 1964.
"Studies on subjective duration. I: Differential sensitivity in the perception of repeated temporal intervals." Acta Psychologica, 22:441-450.

Michon, John A. 1965.
"Studies on subjective time. II: Subjective time measurement during tasks with different information content." Acta Psychologica, 24:205-219.

Michon, John A. 1966.
"Tapping regularity as a measure of perceptual motor load." Ergonomics, 9:401-412.

Michon, John A. 1967.
Timing and Temporal Tracking. Soesterberg, Netherlands: Institute for Perception.

Michon, John A. 1967.
"Magnitude scaling for short duration with closed spaced stimuli." Psychonomic Science, 9:359-360.

Michon, John A. 1972.
"Processing of temporal information and the cognitive theory of time experience." In J.T. Fraser, F.C. Haber, and G.H. Muller (eds.), The Study of Time, pp. 242-258. New York: Springer-Verlag.

Michon, John A. 1974.
"Programs and 'programs' for sequential patterns in motor behaviour." Brain Research, 71:413-424.

Michon, John A. 1975.
"Time experience and memory process." In J.T. Fraser and N. Lawrence (eds.), The Study of Time II, pp. 302-313. New York: Springer-Verlag.

Michon, John A. 1977.
"Holes in the fabric of subjective time: figure ground relations in event sequences." _Acta Psychologica_, 41:191-203.

Michon, John A. 1978.
"The making of the present: a tutorial review." _Attention and Performance_, 7:89-111.

Michon, John A. 1978.
"The perception and evaluation of time from a psychological viewpoint." _Epistenologia_, 1:223-250.

Michon, John A. 1985.
"The compleat time experiencer." In John A. Michon and Janet L. Jackson (eds.), _Time, Mind, and Behavior_, pp. 20-52. New York: Springer-Verlag.

Michon, John A. 1985.
"Temporality and metaphor." In John A. Michon and Janet L. Jackson (eds.), _Time, Mind, and Behavior_, pp. 288-296. New York: Springer-Verlag.

Michon, John A. 1989.
"Timing your mind and minding your time." In J.T. Fraser (ed.), _Time and Mind: The Study of Time VI_. Madison, CT: International Universities Press.

Michon, John A., and Janet L. Jackson. 1984.
"Attentional effort and cognitive strategies in the processing of temporal information." _Annals of the New York Academy of Sciences_, 423:298-321.

Michon, John A., and Janet L. Jackson. 1985.
Time, Mind, and Behavior. New York: Springer-Verlag.

Michon, John A., and Janet L. Jackson. 1985.
"Introduction: the psychology of time." In John A. Michon and Janet L. Jackson (eds.), _Time, Mind, and Behavior_, pp. 2-17. New York: Springer-Verlag.

Michon, John A., Viviane Pouthas, and Janet L. Jackson. 1988.
Two Essays on the Origin of the Idea of Time According to Guyau. Haren: Institute for Experimental Psychology.

Milburn, Michael A. 1978.
"Sources of bias in the prediction of future events." Organizational Behavior and Human Performance, 21(1):17-26.

Miller, Dale T., and Carol A. Porter. 1980.
"Effects of temporal perspective on the attribution process." Journal of Personality and Social Psychology, 39(4):532-541.

Miller, J., and R. Anbar. 1981.
"Expectancy and frequency effects on perceptual and motor systems in choice reaction time." Memory and Cognition, 9:631-641.

Milligan, W.L., D.A. Powell, Caro Harley, and Ernest Fuchtgott. 1984.
"A comparison of physical health and psychosocial variables as predictors of reaction time and serial learning performance in elderly men." Journal of Gerontology, 39(6):704-710.

Minor, Craig, Mark Todorovich, James Boyk, and George P. Moore. 1984.
"A computer-based portable keyboard monitor for studying timing performance in pianists." Annals of the New York Academy of Sciences, 423:651-652.

Mischel, Walter. 1958.
"Preference for delayed reinforcement: an experimental study of a cultural observation." Journal of Abnormal and Social Psychology, 56:57-60.

Mischel, Walter. 1961.
"Delay of gratification, need for achievement, and acquiescence in another culture." Journal of Abnormal and Social Psychology, 62:543-552.

Mischel, Walter, and R. Metzner. 1962.
"Preference for delayed reward as a function of age, intelligence, and length of delay interval." Journal of Abnormal and Social Psychology, 64:425-431.

Mitchell, C. Thomas, and Roy Davis. 1987.
"The perception of time in scale model environments." Perception, 16(1):5-16.

Miyatani, Makoto. 1984-85.
"The time and distance judgments at different levels of discriminability of temporal and spatial information." Hiroshima Forum for Psychology, 10:45-55.

Mo, Suchoon S., and Mark R. Chavez. 1986.
"Perceptual aberration and brain hemisphere reversal of foreperiod effect of time estimation." Journal of Clinical Psychology, 42(5):91-105.

Mogar, M. 1960.
"Children's causal reasoning about natural phenomena." Child Development, 31:59-65.

Mohan, Vidhu, and Rupal Deol. 1983.
"Affect of personality and frequency of KR on acquisition of efficiency in time estimation." Journal of Psychological Researches, 27(2):92-99.

Monks, F. 1968.
"Future time perspective in adolescents." Human Development, 11(2):107-123.

Montanegro, Jacques. 1985.
"The development of temporal inferences and meanings in 5- to 8-year old children." In John A. Michon and Janet L. Jackson (eds.), Time, Mind, and Behavior, pp. 279-287. New York: Springer-Verlag.

Montare, Alberto. 1983.
"Temporal correlates of discrimination learning." Perceptual and Motor Skills, 57(3):763-774.

Montare, Alberto. 1983.
"Temporal correlates of children's discrimination learning." Journal of Psychology, 115(2):137-144.

Montare, Alberto. 1985.
"Learning effects of knowledge of results upon time estimation." Perceptual and Motor Skills, 60(3):871-877.

Montroy, P., D. McManis, and D. Bell. 1971.
"Development of time concepts in normal and retarded children." Psychological Reports, 28:895-902.

Morasso, P., F. Mussa Ivaldi, and C. Ruggiero. 1983.
"How a discontinuous mechanism can produce continuous

patterns in trajectory formation and handwriting." Acta Psychologica, 54:83-96.

Moreland, Richard L., and John M. Levine. 1988. "Group dynamics over time: development and socialization in small groups." In Joseph E. McGrath (ed.), The Social Psychology of Time: New Perspectives, pp. 151-181. Newbury Park, CA: Sage.

Mori, I. 1976. "A cross-cultural study on children's conception of speed and duration: a comparison between Japanese and Thai children." Japanese Psychological Research, 18:105-112.

Moriya, Kunimitsu. 1975. "Attitudes toward the future: a comparative study of younger and older samples." Journal of Child Development, 11:39-44.

Moriya, Kunimitsu. 1979. "Time changes of attitudes toward the future among younger people." Journal of Child Development, 15:33-39.

Morton, J. 1968. "Repeated items and decay in memory." Psychonomic Science, 10:219-220.

Mulligan, R.M., and H.R. Schiffman. 1979. "Temporal experience as a function of organization in memory." Bulletin of the Psychonomic Society, 14:417-420.

Murdock, B.B. 1972. "Short-term memory." In G.H. Bower (ed.), The Psychology of Learning and Motivation: Advances in Research and Theory, Vol. 5, pp. 67-127. New York: Academic.

Murdock, B.B. 1976. "Item and order information in short-term serial memory." Journal of Experimental Psychology: General, 105:191-216.

Murphy, Gardner. 1967. "Direct contacts with past and future: Retrocognition and precognition." Journal of the American Society for Psychical Research, 61(1):3-23.

Murphy, Thomas J., and Alan S. De Wolfe. 1985-86. "Future time perspective in alcoholics, process and reactive schizophrenics, and normals." International Journal of the Addictions, 20(11-12):1815-1822.

Muskatel, Nadine. 1984. "Effect of meditation training on aspects of coronary-prone behavior." Perceptual and Motor Skills, 58(2):515-518.

Mykel, Scott J. 1985. "Discussion of Dr. G.H. Fenchel's paper 'Time as self and ego experience'." Issues in Ego Psychology, 8(1-2):82-86.

Naatanen, Risto, and H. Summala. 1986. "One's advantage and perceptual errors: a demonstration of motivational bias." Journal of Psychology, 120(3):225-228.

Nass, M.L. 1956. "The effects of three variables on children's concepts of physical causality." Journal of Abnormal and Social Psychology, 53:191-196.

Natsoulas, Thomas. 1988. "Gibson, James, and the temporal continuity of experience." Imagination, Cognition and Personality, 7(4):351-376.

Nawas, M. Mike, and Jerome J. Platt. 1965. "A future-oriented theory of nostalgia." Journal of Individual Psychology, 21:51-57.

Nelson, Gordon K. 1983. "Time in developmental studies: a convergence of the dialectic and phenomenological thought." Genetic Psychology Monographs, 108:215-243.

Nelson, T.M., T.H. Nilsson, and M. Johnson. 1984. "Interaction of temperature illuminance and apparent time on sedentary work fatigue." Ergonomics, 27(1):89-101.

Neugarten, Bernice L., and N. Dattan. 1973. "Sociological perspectives on the life cycle." In P.B. Baltes, and K.W. Schaie (eds.), Life Span Developmental

Psychology: Personality and Socialization, pp. 53-71. New York: Academic.

Newell, Stephens. 1972.
"Chemical modifiers of time perception." In Henri Yaker, Humphry Osmond, and Frances Cheek (eds.), The Future of Time: Man's Temporal Environment, pp. 351-388. Garden City, NY: Doubleday.

Newman, Margaret A., and J.K. Gaudiano. 1984.
"Depression as an explanation for decreased subjective time in the elderly." Nursing Research, 33(3):137-139.

Nichols, H. 1891.
"The psychology of time." American Journal of Psychology, 3:453-529.

Niihara, Tomoko, and Katsuhiko Kusano. 1984.
"Temporal accuracy of rhythmic movement in mentally retarded children." Japanese Journal of Educational Psychology, 32(1):18-24.

Nisan, M. 1972.
"Dimension of time in relation to choice behavior and achievement orientation." Journal of Personality and Social Psychology, 21:175-182.

Niville, Ann, Amy Kreisberg, and Gary Kielhofner. 1985.
"Temporal dysfunction in schizophrenia." Occupational Therapy in Mental Health, 5(1):1-19.

Nuttin, Joseph R. 1964.
"The future time perspective in human motivation and learning." Acta Psychologica, 23:60-82.

Nuttin, Joseph R. 1985.
Future Time Perspective and Motivation: Theory and Research Method. Hillsdale, NJ: Erlbaum.

Oakden, E.C., and Mary Sturt. 1922.
"The development of the knowledge of time in children." British Journal of Psychology, 12:309-336.

Olum, V. 1956.
"Developmental differences in the perception of causality." American Journal of Psychology, 69:417-423.

Oppenheimer, Louis, and Huub Van der Lee. 1983. "Social perspective-taking and event sequences: a developmental study." Psychological Reports, 53(3):683-690.

Orme, John E. 1962. "Time estimation and personality." Journal of Mental Science, 108:213-216.

Orme, John E. 1964. "Personality, time estimation and time experience." Acta Psychologica, 22:430-440.

Orme, John E. 1969. Time, Experience and Behavior. New York: Elsevier.

Orme, John E. 1969. "The time location of associations in schizophrenia." Journal of Clinical Psychology, 25:260-261.

Orme, John E. 1978. "Time: psychological aspects." In Tommy Carlstein, Don Parkes, and Nigel Thrift (eds.), Making Sense of Time, pp. 66-75. New York: Wiley.

Ornstein, Robert E. 1970. On the Experience of Time. Baltimore, MD: Penguin.

Osberg, Timothy, and Sidney J. Shrauger. 1986. "Restropective versus prospective causal judgments of self and others' behavior." Journal of Social Psychology, 126(2):169-178.

Paivio, A. 1978. "Comparison of mental clocks." Journal of Experimental Psychology: Human Perception and Performance, 4:61-71.

Panides, Wallace. 1984. "The perception of the past, present, and future in preadolescent asthmatic children: an exploratory study. Sex Roles, 11(11-12):1141-1152.

Park, David. 1985. "Brain time and mind time." In John A. Michon and Janet L. Jackson (eds.), Time, Mind, and Behavior, pp. 53-64. New York: Springer-Verlag.

Pena-Correal, Telmo E., and A.W. Logue. 1984.
"Self-control and responding during reinforcement delay." _Annals of the New York Academy of Sciences_, 423:618-621.

Pendse, Shripad G. 1979.
"Perception: an experiment to compare 'here and now' and 'there and then' perspectives." _Perceptual and Motor Skills_, 48(3,Pt.2):1215-1218.

Peters, Michael. 1981.
"Attentional asymmetries during concurrent bimanual performance." _Quarterly Journal of Experimental Psychology_, 33A:95-103.

Peters, Michael. 1984.
"Timing of initiation and termination of dual manual movements." _Annals of the New York Academy of Sciences_, 423:628-629.

Peterson, L.R., S.T. Johnson, and R. Coatney. 1969.
"The effect of repeated occurrences on judgments of recency." _Journal of Verbal Learning and Verbal Behavior_, 8:591-596.

Peterson-Cooney, Lorrie. 1987.
"Time concentrated instruction as an immediate risk to self-actualization." _Psychological Reports_, 61(1):183-190.

Petrusic, William M. 1984.
"Explicit counting and time-order errors in duration discrimination." _Annals of the New York Academy of Sciences_, 423:630-633.

Pettit, T.F. 1969.
"Anality and time." _Journal of Consulting and Clinical Psychology_, 33:170-174.

Pfaff, D. 1968.
"Effects of temperature and time of day on time judgments." _Journal of Experimental Psychology_, 76:419-422.

Piaget, Jean. 1970.
The Child's Conception of Movement and Speed. New York: Ballantine.

Piaget, Jean. 1970.
The Child's Conception of Time. New York: Basic Books.

Piaget, Jean. 1981.
"Time perception in children." In J.T. Fraser (ed.), The Voices of Time, 2nd edn., pp. 202-216. Amherst, MA: University of Massachusetts Press.

Platt, Jerome J., and Russell Eisenman. 1968.
"Internal-external control of reinforcement, time perspective adjustment and anxiety." Journal of General Psychology, 79(1):121-128.

Platt, Jerome J., Russell Eisenman, O. De Lisser, and A. Darbes. 1971.
"Temporal perspective as a personality dimension in college students: a re-evaluation." Perceptual and Motor Skills, 33:103-109.

Platt, Jerome J., Russel Eisenman, and Edward De Gross. 1969.
"Birth order and sex differences in future time perspective." Developmental Psychology, 1(1):70.

Platt, Jerome. J., and Robert E. Taylor. 1967.
"Homesickness, future time perspective, and the self concept." Journal of Individual Psychology, 23:94-97.

Platt, John R. 1979.
"Interresponse-time shaping by variable-interval-like interresponse-time reinforcement contingencies." Journal of Experimental Analysis of Behavior, 31:3-14.

Platt, John R. 1979.
"Temporal differentiation and the psychophysics of time." In M.D. Zeiler and P. Harzem (eds.), Advances in the Analysis of Behavior: Reinforcement and the Organization of Behavior, pp. 1-29. New York: Wiley.

Platt, John R. 1984.
"Motivational and response factors in temporal differentiation." Annals of the New York Academy of Sciences, 423:200-210.

Platt, John R., and E.R. Davis. 1983.
"Bisection of temporal intervals by pigeons." Journal of

Experimental Psychology: Animal Behavior Processes, 9:160-170.

Poppel, Ernst. 1972.
"Oscillations as possible basis for time perception." In J.T. Fraser, F.C. Haber, and G.H. Muller (eds.), The Study of Time, pp. 219-241. New York: Springer-Verlag.

Poppel, Ernst. 1976.
"Time perception." In H.L. Tenber, R. Held, and H. Leibowitz (eds.), Handbook of Sensory Physiology, Vol. 8, pp. 713-729. New York: Springer-Verlag.

Poppel, Ernst. 1987.
"Time perception." In George Adelman (ed.), Encyclopedia of Neuroscience, 2nd edn., pp. 1215-1216. Boston. MA: Birkhauser.

Poulton, E.C. 1957.
"On prediction in skilled movements." Psychological Bulletin, 54:467-478.

Pouthas, Viviane. 1985.
"Timing behavior in young children: a developmental approach to conditioned spaced responding." In John A. Michon and Janet L. Jackson (eds.), Time, Mind, and Behavior, pp. 100-109. New York: Springer-Verlag.

Pouthas, Viviane, and Anne-Yvonne Jacquet. 1987.
"A developmental study of timing behavior in 4 1/2 and 7-year old children." Journal of Experimental Child Psychology, 43(2):282-299.

Povel, Dirk-Jan. 1981.
"Internal representation of simple temporal patterns." Journal of Experimental Psychology: Human Perception and Performance, 7:3-18.

Povel, Dirk-Jan. 1984.
"A theoretical framework for rhythm perception." Psychological Research, 45:315-337.

Povel, Dirk-Jan. 1985.
"Time, rhythms and tension: in search of the determinants of rhythmicity." In John A. Michon and Janet L. Jackson (eds.), Time, Mind, and Behavior, pp. 215-225. New York: Springer-Verlag.

Povel, Dirk-Jan, and R. Collard. 1982.
"Structural factors in patterned finger tapping." _Acta Psychologica_, 52:107-123.

Povel, Dirk-Jan, and H. Okkerman. 1981.
"Accents in equitone sequences." _Perception and Psychophysics_, 30:565-572.

Poynter, W.D. 1983.
"Duration judgment and the segmentation of experience." _Memory and Cognition_, 11:77-82.

Poynter, W.D., and D. Homa. 1983.
"Duration judgment and the experience of change." _Perception and Psychophysics_, 33:548-560.

Predebon, John. 1984.
"Organization of stimulus events and remembered apparent duration." _Australian Journal of PSychology_, 36(2):161-169.

Predebon, John. 1988.
"Retrospective time judgments and clock duration." _Perceptual and Motor Skills_, 66(1):19-24.

Price, J.R., and W. Tare. 1975.
"A cross-cultural study of recall of time-related and non-time-related verbal material." _International Journal of Psychology_, 10:247-254.

Quigley, John I., A.L. Combs, and N. O'Leary. 1984.
"Sensed duration of time: influence of time as a barrier." _Perceptual and Motor Skills_, 58(1):72-74.

Rabin, Albert I. 1978.
"Future time perspective and ego strength." In J.T. Fraser, N. Lawrence, and D. Park (eds.), _The Study of Time III_, pp. 294-305. New York: Springer-Verlag.

Rachlin, H.C., and L. Green. 1972.
"Commitment, choice and self-control." _Journal of Experimental Analysis of Behavior_, 17:15-22.

Rai, S.N. 1973.
"Time estimation by reproduction method as a function of auditory distraction." _Psychological Studies_, 1:59-65.

Rakowski, William. 1979.
"Future time perspectives in later adulthood: review and research directions." Experimental Aging Research, 5(1):43-88.

Rakowski, William. 1982.
"Comparability and reliability of techniques to assess personal future extension." Experimental Aging Research, 8(3-4):199-201.

Rakowski, William. 1982.
"Temporal context and the perceptions of geriatric patients." Social Science and Medicine, 16(3):241-244.

Rakowski, William. 1986.
"Future time perspective: applications to the health context of later adulthood." American Behavioral Scientist, 29(6):730-745.

Rakowski, William, Clifton E. Barber, and Wayne C. Seelbach. 1980.
"Reponse format and coding procedures in research on time perspective: a methodological commentary." Perceptual and Motor Skills, 51(1):39-48.

Rakowski, Willian, Cliton E. Barber, and Wayne C. Seelbach. 1980.
"Comparability and short-term stability of techniques to assess extension of thinking about one's personal future." Perceptual and Motor Skills, 56(2):559-564.

Ramella, Reno J. 1983.
"Processing, knowledge of results and a multi-dimensional task." Perceptual and Motor Skills, 57(1):43-48.

Rao, P. Rama. 1978.
Studies in Time Perception. Delhi, India: Concept Publishing.

Rappaport, Herbert, Kathy Enrich, and Arnold Wilson. 1985.
"Relation between ego identity and temporal perspective." Journal of Personality and Social Psychology, 48(6):1609-1620.

Raslear, Thomas G. 1985.
"Perceptual bias and response bias in temporal bisection." Perception and Psychophysics, 38(3):261-268.

Raundalen, Magne, and O.J. Finney. 1986.
"Children's and teen-agers' views of the future." Journal of Mental Health, 15(1-3):114-125.

Raynor, Joel O., and Elliot E. Entin. 1983.
"The function of future orientation as a determinant of human behavior in step-path theory of action." International Journal of Psychology, 18(5):463-487.

Reichenbach, M., and A. Mathern. 1959.
"The place of time and aging in the natural sciences and scientific philosophy." In J.E. Birren (ed.), Handbook of Aging and the Individual, pp. 43-80. Chicago, IL: University of Chicago Press.

Restle, F. 1970.
"Theory of serial pattern learning: structural trees." Psychological Review, 77:481-495.

Restle, F. 1972.
"Serial patterns: the role of phrasing." Journal of Experimental Psychology, 92:385-390.

Restle, F. 1976.
"Structural ambiguity in serial pattern learning." Cognitive Psychology, 8:357-381.

Restle, F., and E.R. Brown. 1970.
"Serial pattern learning." Journal of Experimental Psychology, 83:120-125.

Restle, F., and E.R. Brown. 1970.
"Organization of serial pattern learning." In G.H. Bower (ed.), The Psychology of Learning and Motivation: Advances in Research and Theory, Vol. 4, pp. 249-331. New York: Academic.

Richards, D. Dean. 1982.
"Children's time concepts: going the distance." In William J. Friedman (ed.), The Developmental Psychology of Time, pp. 13-45. New York: Academic.

Richards, W. 1973.
"Time reproductions by H.M." Acta Psychologica, 37:279-282.

Richelle, Marc, and Helga Lejeune. 1980.
Time in Animal Behavior. New York: Pergamon.

Richelle, Marc, and Helga Lejeune. 1984.
"Timing competence and timing performance: A cross-species approach." Annals of the New York Academy of Sciences, 423:254-268.

Richelle, Marc, Helga Lejeune, Jean-Jacques Perikel, and Patrik Fery. 1985.
"From biotemporality to nontemporality: toward an integrative and comparative view of time in behavior." In John A. Michon and Janet L. Jackson (eds.), Time, Mind, and Behavior, pp. 75-99. New York: Springer-Verlag.

Ricks, D., C. Umberger, and R.A. Mack. 1964.
"A measure of increased temporal perspective in successfully treated adolescent delinquent boys." Journal of Abnormal and Social Psychology, 69:685-689.

Riegel, Klaus F. 1972.
"Time and change in the development of the individual and society." In H. Reese (ed.), Advances in Child Development and Behavior, Vol. 7, pp. 81-113. New York: Academic.

Riegel, Klaus F. 1973.
"The recall of historical events." Behavioral Science, 18:354-363.

Riegel, Klaus F. 1977.
"The dialectics of time." In N. Datan and H.W. Reese (eds.), Life-Span Developmental Psychology: Dialectical Perspectives on Experimental Research, pp. 3-45. New York: Academic.

Riegel, Klaus F. 1977.
"Toward a dialectical interpretation of time and change." In Bernard S. Gorman and Alden E. Wessman (eds.), The Personal Experience of Time, pp. 59-108. New York: Plenum.

Rimoldi, Horacio J.A. 1951.
"Personal tempo." Journal of Abnormal and Social Psychology, 46:283-303.

Rimoldi, Horacio J.A., and S. Cabanski. 1961.
"Temporal organization of behavior." Journal of Psychology, 51:383-91.

Riss, Walter. 1984.
"Testing a theory of brain function by computer methods: IV. Comparing spatial and temporal order in the two cerebral hemispheres." Brain, Behaviour and Evolution, 24(4):177-183.

Robbins, Michael, et al. 1968.
"Time estimation and acculturation among the Baganda." Perceptual and Motor Skills, 26:1010.

Roberts, Alan H., and Joel E. Greene. 1971.
"Cross-cultural study of relationships among four dimensions of time perspective." Perceptual and Motor Skills, 33(1):163-173.

Roberts, Alan H., and Robert S. Herrman. 1960.
"Dogmatism, time perspective, and anomie." Journal of Individual Psychology, 16:67-72.

Robertson, S.A. 1978.
"Some personality correlates of time competance, temporal extension and temporal evaluation." Perceptual and Motor Skills, 46(3):743-750.

Roby, Thornton B., and Jeffrey Z. Rubin. 1973.
"An exploratory study of competitive temporal judgement." Behaviorial Science, 18(1):42-51.

Roelofs, C.O., and W.P.C. Zeaman. 1951.
"Influence of different sequences of optical stimuli on the estimation of the duration of a given interval of time." Acta Psychologica, 8:89-128.

Roessler, Richard T., and Steven E. Boone. 1977.
"Life outlook of rehabilitation clients: a comparative analysis." Journal of Applied Rehabilitation Counseling, 8(2):89-98.

Rokeach, Milton, and R. Bonier. 1960.
"Time perspective, dogmatism, and anxiety." In Milton Rokeach (ed.), The Open and Closed Mind, pp. 366-375. New York: Basic Books.

Roos, P., and R.J. Albers. 1965.
"Performance of retardates and normals on a measure of temporal orientation." American Journal of Mental Psychology, 69:835-838.

Rose, Robert J. 1984.
"Processsing time for repetitions and the spacing effect." Canadian Journal of Psychology, 38(4):537-550.

Rosenthal, Leslie, and Margot C. Howe. 1984.
"A comparison of temporal adaptation among day versus night shift workers." Occupational Therapy in Mental Health, 4(2):59-78.

Rosenbaum, D.A., S.B. Kenny, and M.A. Derr. 1983.
"Hierarchical control to rapid movement sequences." Journal of Experimental Psychology: Human Perception and Performance, 9:86-102.

Rousseau, Robert, and Daniel Picard. 1984.
"An adaptive counter model for time estimation." Annals of the New York Academy of Sciences, 423:639-642.

Rousseau, Robert, Jocelyn Poirier, and Gerard Tremblay. 1984.
"Temporal organization and intermodality in duration discrimination of short empty intervals." Annals of the New York Academu of Sciences, 423:636-637.

Rozenzweig, S., and A.C. Koht. 1933.
"The experience of duration as affected by need tension." Journal of Experimental Psychology, 16:745-755.

Ruggiere, V., and A. Guerrera. 1982.
"Time estimation in some group situations." Perceptual and Motor Skills, 54:31-39.

Ruiz, Rene A., Ronald S. Reivich, and Herbert H. Krauss. 1967.
"Tests of temporal perspective: do they measure the same construct?" Psychological Reports, 21(3):849-852.

Rule, Stanley I. 1983.
"Composition rule for perceived duration of simultaneous events." Perception and Psychophysics, 34(6):569-572.

Rule, Stanley J., and Dwight Curtis. 1985. "Ordinal properties of perceived average duration: simultaneous and sequential presentations." Journal of Experimental Psychology: Human Perception and Performance, 11(4):509-516.

Rumelhart, D.E., and D.A. Norman. 1982. "Simulating a skilled typist: a study of skilled cognitive-motor performance." Cognitive Science, 6:1-36.

Rumiati, Rino, and Sergio Roncato. 1985. "Time representation in sentence comprehension." Acta Psychologica, 59(3):271-283.

Rutschmann, R. 1973. "Visual perception of temporal order." In S. Kornblum (ed.), Attention and Performance IV, pp. 687-701. New York: Academic.

Rychlak, J.F. 1972. "Manifest anxiety as reflecting commitment to the psychological present at the expense of cognitive futurity." Journal of Consulting and Clinical Psychology, 38:70-79.

Sacher, George A. 1967. "The complementarity of entropy terms for the temperature-dependence of development and aging." Annals of the New York Academy of Sciences, 138:680-712.

Salzinger, Kurt. 1972. "The immediacy hypothesis of schizophrenia." In Henri Yaker, Humphry Osmond, and Frances Cheek (eds.), The Future of Time: Man's Temporal Enviroment, pp. 272-292. Garden City, NY: Doubleday.

Samuelson, Paul Anthony. 1976. "Speeding up of time with age in recognition of life as fleeting." In Anthony M. Tang, Fred M. Westfield, and James S. Worley (eds.), Evolution, Welfare, and Time in Economics: Essays in Honor of Nicholas Georgescu-Roegen, pp. 153-170. Lexington, MA: D.C. Heath

Sanders, Jeffrey L. 1985. "Connotiative meaning of time among Egyptian and American students." Journal of Social Psychology, 125(5):587-590.

Sanders, Sharon A. 1986.
"Development of a tool to measure subjective time experience." Nursing Research, 35(3):178-182.

Saunders, David R. 1985.
"The internal structure of time estimation task." Perceptual and Motor Skills, 60(2):618.

Saunders, D.S. 1976.
Insect Clocks. Oxford, UK: Pergamon.

Savickas, Mark L., S. Marc Silling, and Steven Schwartz. 1984.
"Time perspective in vocational maturity and career decision making." Journal of Vocational Behavior, 25(3):258-269.

Schab, Frank R., and Robert G. Crowder. 1988.
"The role of succession in temporal cognition: is the time-order error a recency effect of memory?" Perception and Psychophysics, 44(3):233-242.

Schachter, S., and L.P. Gross. 1968.
"Manipulated time and eating behavior." Journal of Personality and Social Psychology, 10:98-106.

Schaie, K. Warner. 1986.
"Beyond calendar definitions of age, time and cohort: the general developmental model revisited." Developmental Review, 6(3):252-277.

Schiffman, H.R., and D.J. Bobko. 1974.
"Effects of stimulus complexity on the perception of brief temporal intervals." Journal of Experimental Psychology, 103:156-159.

Schiffman, H.R., and D.J. Bobko. 1977.
"The role of number and familiarity of stimuli in the perception of brief temporal intervals." American Journal of Psychology, 90:85-93.

Schilder, P. 1936.
"Psychopathology of time." Journal of Nervous and Mental Disorders, 83:530-546.

Schmeidler, Gertrude R. 1964.
"An experiment on precognitive clairvoyance. Part III:

Precognition scores related to the subjects' ways of viewing time." Journal of Psychology, 28:93-101.

Schmidt, R.A. 1980.
"On the theoretical status of time in motor-program representations." In G. Stelmach and J. Requin (eds.), Tutorials in Motor Behavior. Amsterdam, Netherlands: North Holland.

Schmidt, R.A., and R.W. Christina. 1969.
"Proprioception as a mediator in the timing of motor responses." Journal of Experimental Psychology, 81:303-307.

Schmidt, Rolf W., Helmut Lamm, and Gisela Trommsdorff. 1978.
"Social class and sex as determinants of future orientation (time perspective) in adults." European Journal of Social Psychology, 8(1):71-90.

Schulz, D. 1963.
"Time, awareness, and the order of presentation in opinion change," Journal of Applied Psychology, 47:280-283.

Schweickert, Richard J. 1978.
"A critical path generalization of the additive factor method: analysis of a Stroop task." Journal of Mathematical Psychology, 18:105-139.

Schweickert, Richard J. 1984.
"The representation of mental activities in critical path networks." Annals of the New York Academy of Sciences, 423:82-95.

Scott, L.H. 1935.
"Time-order errors in the discrimination of short tonal duration." Journal of Experimental Psychology, 18:741-766.

Sebel, Aliza I., and W.E. Wilsoncroft. 1983.
"Auditory and visual differences in time perception." Perceptual and Motor Skills, 57(1):295-300.

Sechrest, L. 1963.
"The psychology of personal constructs: George Kelly."

In Joseph M. Wepman and Ralph W. Heine (eds.) , Concepts of Personality, pp. 206-233. Chicago: Aldine.

Semjen, Andras, Adela Garcia-Colera, and Jean Requin. 1984. "On controlling force and time in rhythmic movement sequences: the effect of stress location." Annals of the New York Academy of Sciences, 423:168-182.

Seth, Usha. 1986. "Future time orientation of boys and girls of different educational faculties." Indian Journal of Current Psychological Research, 1(2):107-111.

Sevriere, J., C. Micelli, and Y. Galifret. 1977. "A psychophysical study of the visual perception of 'instantaneous' and 'durable'." Vision Research, 17:57-63.

Shaffer, G., and A.R. Gilliland. 1938. "The relation of time estimation to certain psychological changes." Journal of Experimental Psychology, 23:545-552.

Shaffer, L. Henry. 1973. "Latency mechanisms in transcription." In S. Kornblum (ed.), Attention and Performance IV, pp. 17-41. New York: Academic.

Shaffer, L. Henry. 1978. "Timing in the motor programming of typing." Quarterly Journal of Experimental Psychology, 30:333-345.

Shaffer, L. Henry. 1981. "Performances of Chopin, Bach, and Bartok: Studies in motor programming." Cognitive Psychology, 13:327-376.

Shaffer, L. Henry. 1982. "Rhythm and timing in skill." Psychological Review, 89:109-122.

Shaffer, L. Henry. 1984. "Timing in solo and duet piano performances." Quarterly Journal of Experimental Psychology, 36A:577-595.

Shaffer, L. Henry. 1985. "Timing in action." In John A. Michon and Janet L.

Jackson (eds.), _Time, Mind, and Behavior_, pp. 226-241. New York: Springer-Verlag.

Shannon, Lael. 1975.
"Development of time perspective in three cultural groups: a cultural difference or an expectancy interpretation." _Developmental Psychology_, 11(1):114-115.

Sherwood, Stephen L. 1967.
"The abandonment of the absolute." _Annals of the New York Academy of Sciences_, 138:600-615.

Shifflett, Peggy A. 1987.
"Future time perspective, past experiences, and negotiation of food use patterns among the aged." _Gerontologist_, 27(5):611-615.

Shifflett, Peggy A., and William A. McIntosh. 1986.
"Food habits and future time: an exploratory study of age appropriate food habits among the elderly." _International Journal of Aging and Human Development_, 24(1):1-17.

Shiffrin, R.M., and J.R. Cook. 1978.
"Short-term forgetting of item and order information." _Journal of Verbal Learning and Verbal Behavior_, 17:189-218.

Shimp, Charles P. 1968.
"Magnitude and frequency of reinforcement and frequencies of interresponse times." _Journal of the Experimental Analysis of Behavior_, 11:525-535.

Shimp, Charles P. 1969.
"Concurrent reinforcement of two interresponse times: the relative frequency of an interresponse time equals its relative harmonic length." _Journal of the Experimental Analysis of Behavior_, 12:403-411.

Shimp, Charles P. 1970.
"Concurrent reinforcement of two interresponse times: absolute rate of reinforcement." _Journal of the Experimental Analysis of Behavior_, 13:1-8.

Shimp, Charles P. 1978.
"Memory, temporal discrimination and learned structure of

behavior." In G.H. Bower (ed.), The Psychology of Learning and Motivation, Vol. 12, pp. 39-76. New York: Academic.

Shimp, Charles P. 1984.
"Timing, learning, and forgetting." Annals of the New York Academy of Sciences, 423:346-360.

Shostrom, Everett L. 1968.
"Time as an integrating factor." In Charlotte Buhler and Fred Massarik (eds.), The Course of Human Life, pp. 351-359. New York: Springer-Verlag.

Shultz, T., and R. Mendelson. 1975.
"The use of covariation as a principle of causal analysis." Child Development, 46:394-399.

Shybut, John. 1970.
"Internal vs. external control, time perspective and delay of gratification of high and low ego strength groups." Journal of Clinical Psychology, 26(4):430-431.

Siegel, Stephen F. 1986.
"A test of the similarity rule model of temporal bisection." Learning and Motivation, 7(1):59-75.

Siegel, Stephen F., and Russell M. Church. 1984.
"The decision rule in temporal bisection." Annals of the New York Academy of Sciences, 423:643-645.

Siegler, R.S. 1975.
"Defining the locus of developmental differences in children's causal reasoning." Journal of Experimental Child Psychology, 20:512-525.

Siegler, R.S., and R. Liebert. 1974.
"Effects of contiguity, regularity and age on children's causal inferences." Developmental Psychology, 10:574-579.

Siegler, R.S., and D. Dean Richards. 1979.
"The development of time, speed and distance concepts." Developmental Psychology, 15:288-296.

Siegman, Aron Wolfe. 1961.
"The relationship between future time perspective, time estimation, and impulse control in a group of young

offenders and in a control group." Journal of Consulting Psychology, 25:470-475.

Siegman. Aron Wolfe. 1962.
"Anxiety, impulse control, intelligence, and the estimation of time." Journal of Clinical Psychology, 18:103-105.

Siegman. Aron Wolfe. 1962.
"Future-time perspective and the perception of duration." Perceptual and Motor Skills, 15:609-610.

Silber, Thomas J. 1983.
"The art of healing and the many worlds of time." Journal of Religion and Health, 22(2):105-115.

Silver, Rae, and Eric L. Bittman. 1984.
"Reproductive mechanisms: interaction of circadian and interval timing." Annals of the New York Academy of Sciences, 423:488-514.

Silverstein, C.I., and A.M. Owens. 1982.
"Perception of time during examination." Perceptual and Motor Skills, 55:298.

Simon, H.A. 1972.
"Complexity and the representation of patterned sequences of symbols." Psychological Review, 79:369-382.

Simon, H.A. and K. Kotovsky. 1963.
"Human acquisition of concepts for sequential patterns." Psychological Review, 70:534-546.

Singer, Jerome L., Harold Wilensky, and Vivian G. McCraven. 1956.
"Delaying capacity, fantasy, and planning ability: a factorial study of some basic ego functions." Journal of Consulting Psychology, 20"378-383.

Skelly, June J., Anthony Rizzuto, and Glenn Wilson. 1984.
"Temporal patterning and selective effects of the human evoked response." Annals of the New York Academy of Sciences, 423:646-648.

Slife, Brent D. 1981.
"Psychology's reliance on linear time: a reformation." Journal of Mind and Behavior, 2(1):27-46.

Small, Jr., A.M., and R.A. Campbell. 1962.
"Temporal differential sensitivity for auditory stimuli." American Journal of Psychology, 75:401-410.

Small, Mark A., J.F. Raney, Terry J. Knapp. 1987.
"Comparision of two reaction time tasks and their relation to intelligence." Perceptual and Motor Skills, 56(3):867-870.

Smirnov, A.G., and M.V. Polyakova. 1984.
"Assessment of a time interval by the different fingers." Human Psychology, 9(3):156-161.

Smith, C. 1980.
"The acquisition of time talk: relations between child and adult grammars." Journal of Child Language, 7:263-278.

Smith, P. Hull. 1984.
"Five-month-old infant recall and utilization of temporal organization." Journal of Experimental Child Psychology, 38(3):400-414.

Smith, R.N., and P. Tomlinson. 1977.
"The development of children's construction of historical duration: a new approach and some findings." Educational Research, 19(3):163-170.

Smythe, E.J., and Sanford Goldstone. 1957.
"The time sense: a normative genetic study of the development of time perception." Perceptual and Motor Skills, 7:49-59.

Sperling, G. 1967.
"Successive approximations to a model for short-term memory." Acta Psychologica, 27:285-292.

Springer, D. 1952.
"Development in young children of an understanding of time and the clock." Journal of Genetic Psychology, 80:83-96.

Squire, L.R., P.M. Chace, and P.C. Slater. 1975.
"Assessment of memory for remote events." Psychological Reports, 37:223-234.

Squyres, Elton M. 1981.
"Guidelines for use in scoring TAT stories for time-span." Perceptual and Motor Skills, 52(1):333-334.

Squyres, Elton M., and Ray A. Craddick. 1982.
"A measure of time perspective with the TAT and some issues of reliability." Journal of Personality Assessment, 46(3):257-259.

Squyres, Elton M., and Ray A. Craddick, K. Burge-Callaway, and B. Kempler. 1982.
"Sex differences in time orientation: a Jungian perspective." Journal of Analytical Psychology, 27(1):71-81.

Staddon, J.E.R. 1974.
"Temporal control, attention and memory." Psychological Review, 81:375-391.

Staddon, J.E.R. 1975.
"Limitations on temporal control: generalization and the effects of context." British Journal of Psychology, 66:229-246.

Staddon, J.E.R. 1984.
"Time and memory." Annals of the New York Academy of Sciences, 423:322-334.

Stark, Stanley. 1962.
"Temporal and atemporal foresight." Journal of Humanistic Psychology, 2(1):56-74.

Stein, Kenneth B., Theodore Sarbin, and James A. Kulik. 1968.
"Future time perspective: its relation to the socialization process and the delinquent role." Journal of Consulting and Clinical Psychology, 32(3):257-264.

Stein, Nancy L., and Christine G. Glenn. 1982.
"Children's concept of time: the development of a story schema." In William J. Friedman (ed.), The Developmental Psychology of Time, pp. 255-282. New York: Academic.

Stein, N., and R. Landis. 1978.
"Effects of age and collateral behavior on temporally discriminated performance of children." Perceptual and Motor Skills, 47:87-94.

Stelmach, George E., and Hans-Leo Teulings. 1983.
"Response characteristics of prepared and restructured handwriting." Acta Psychologica, 54:51-67.

Stelmach, George E., Patricia A. Mullins, and Hans-Leo Teulings. 1983.
"Motor programming and temporal patterns in handwriting." Annals of the New York Academy of Sciences, 423:144-157.

Stephenson, William. 1988.
"William James, Niels Bohr, and complementarity: IV. The significance of time." Psychological Record, 38(1):19-35.

Stern, E.M. 1977.
"Narcissism and the defiance of time." In M.C. Nelson (ed.), The Narcissistic Condition. New York: Human Sciences Press.

Sternberg. Saul, and Ronald L. Knoll. 1973.
"The perception of temporal order: fundamental issues and a general model." In S. Kornblum (ed.), Attention and Performance IV, pp. 629-685. New York: Academic.

Stevens, S.S. 1971.
"Issues in psychophysical measurement." Psychological Review, 78:426-450.

St. Jean, Richard, and Lynda Robertson. 1986.
"Attentional versus absorbtive processing in hypnotic time estimation." Journal of Abnormal Psychology, 95(1):40-42.

Stokes, Douglas M. 1985.
"Theories of anomalous temporal phenomena." Parapsychology Review, 16(6):12-15.

Stolerman, I.P. 1985.
"Cue location does not account for all selective associations in food aversion learning." Quarterly Journal of Experimental Psychology: Comparative and Physiological Psychology, 37(2B):181-185.

Strack, Fritz, et al. 1982.
"Effects of salience and time pressure ratings of social causality." Journal of Experimental Social Psychology, 18:581-594.

Strack, Fritz, Norbert Schwarz, and Elisabeth G. Schneidinger. 1985. "Happiness and reminiscing: the role of time perspective, affect, and mode of thinking." Journal of Personality and Social Psychology, 49(6):1460-1469.

Straus, Erwin (Walter Maximilian). 1982. Man, Time, and World: Two Contributions to Anthropological Psychology. Pittsburgh, PA: Duquesne University Press.

Strickland, Lloyd H., Roy J. Lewicki, and Arnold M. Katz. 1966. "Temporal oreintation and perceived control as determinants of risk-taking." Journal of Experimental Social Psychology, 2:143-151.

Stroud, John M. 1956. "The fine structure of psychological time." In Henry Quastler (ed.), Information Theory in Psychology: Problems and Methods, pp. 174-207. Glencoe, IL: Free Press.

Stroud, John M. 1967. "The fine structure of psychological time." Annals of the New York Academy of Sciences, 138:623-631.

Strube, Michael J., and Cynthia L. Lott. 1984. "Time urgency and type A behavior pattern: implications for time investment and psychological entrapment." Journal of Research in Personality, 18(3):395-409.

Strumpf, Neville E. 1987. "Probing the temporal world of the elderly." International Journal of Nursing Studies, 24(3):210-214.

Strumwasser, F. 1965. "The demonstration and manipulation of a circadian rhythm in a single neuron." In J. Aschoff (ed.), Circadian Clocks, pp. 442-462. Amsterdam, Netherlands: North Holland.

Stubbs, D. Alan. 1979. "Temporal discrimination and psychophysics." In M.D. Zeiler and P. Harzem (eds.), Advances in the Analysis of Behavior: Reinforcement and the Organization of Behavior, Vol. 1, pp. 341-369. Chichester, UK: Wiley.

Stubbs, D. Alan. 1980.
"Temporal discrimination and a free-operant psychophysical procedure." Journal of the Experimental Analysis of Behavior, 33:167-185.

Stubbs, D. Alan, Leon R. Dryfus, and J.Gregor Fetterman. 1984.
"The perception of temporal events." Annals of the New York Academy of Sciences, 423:30-42.

Stubbs, D. Alan, S.J. Vautin, H.M. Reid, and D.L. Delehanty. 1978.
"Discriminative functions of schedule stimuli and memory: a combination of schedule and choice procedures." Journal of the Experimental Analysis of Behavior, 29:167-180.

Sturt, Mary. 1925.
The Psychology of Time. New York: Harcourt, Brace.

Sudnow, David. 1972.
"Temporal parameters of interpersonal observation." In David Sudnow (ed.), Studies in Social Interaction, pp. 259-279. New York: Free Press.

Sundberg, Norman D., Millicent E. Poole, and Leona E. Tyler. 1983.
"Adolescents' expectations of future events: a cross-cultural study of Australians, Americans, and Indians." International Journal of Psychology, 18(5):415-427.

Surwillo, W.W. 1966.
"Time perception and the 'internal clock': some observations on the role of the electroencephalogram." Brain Research, 2:390-392.

Svenson, Ola, and Anne Edland. 1987.
"Change of preferences under time pressure: choices and judgments." Scandinavian Journal of Psychology, 28(4):322-330.

Swindle, P.F. 1913.
"On the inheritance of rhythm." American Journal of Psychology, 24:180-203.

Takahashi, J.S., and M. Zatz. 1982.
"Regulation of circadian rhythmicity." Science, 217:1104-1111.

Teahan, John E. 1958.
"Future time perspective, optimism and academic achievement." Journal of Abnormal and Social Psychology, 57:379-380.

Terman, Michael. 1983.
"Behavioral analysis and circadian rhythms." In M.D. Zeiler and P. Harzem (eds.), Advances in Analysis of Behaviour, Vol. 3, pp. 103-141. Chichester, UK: Wiley.

Terman, Michael, John Gibbon, Stephen Fairhurst, and Amy Waring. 1984.
"Daily meal anticipation: interaction of circadian and interval timing." Annals of the New York Academy of Sciences, 423:470-487.

Terzuolo, C., and P. Viviani. 1980.
"Determinants and characteristics of motor patterns used for typing." Neuroscience, 5:1085-1103.

Teulings, Hans-Leo, and F. Maarse. 1984.
"Digital recording and processing of handwriting movements." Human Movements Science, 3:193-217.

Thayer, S., Bernard S. Gorman, Alden E. Wessman, G. Schmeidler, and E. Mannucci. 1975.
"The relationship between locus of control and temporal experience." Journal of Genetic Psychology, 110:275-279

Theios, J. 1973.
"Reaction time measurement in the study of memory processes: theory and data." In G.H. Bower (ed.), The Psychology of Learning and Motivation: Advances in Research and Theory, Vol 7., pp. 43-85. New York: Academic.

Thomae, Hans. 1981.
"Future time perspective and the problem of cognition/motivation interaction." In Gery d'Ydewalle and Willy Lens (eds.), Cognition in Human Motivation and Learning, 261-274. Hillsdale, NJ: Erlbaum.

Thomas, E.A.C, and I.B. Brown. 1974.
"Time perception and the filled duration illusion." Perception and Psychophysics, 16:449-458.

Thomas, E.A.C., and N.E. Cantor. 1978.
"Interdependence between the processing of temporal and nontemporal information." In J. Requin (ed.), Attention and Performance VIII, pp. 43-62. Hillsdale, NJ: Erlbaum.

Thomas, E.A.C., and W.B. Weaver. 1975.
"Cognitive processing and time perception." Perception and Psychophysics, 17:363-367.

Thomassen, Arnold J.W.M., and Hans-Leo Teulings. 1985.
"Time, size and shape in handwriting: exploring spatio-temporal relationships at different levels." In John A. Michon and Janet L. Jackson (eds.), Time, Mind, and Behavior, pp. 253-263. New York: Springer-Verlag.

Thompson, J.G., H.R. Schiffman, and D.J. Bobko. 1976.
"The discrimination of brief temporal intervals." Acta Psychologica, 40:489-493.

Thor, Donald H. 1962.
"Diurnal variability in time estimation." Perceptual and Motor Skills, 15:451-454.

Thorpe, J.E. 1978.
Rhythmic Activity of Fishes. New York: Academic.

Tismer, K.G. 1985.
"Sex and age differences in personal and global future time perspective: a replication." Perceptual and Motor Skills, 61:1007-1010.

Tismer, K.G. 1987.
"Psychological aspects of temporal dominance during adolescence." Psychological Reports, 61(2):647-654.

Toban, Eileen V. 1968.
"Factors influencing decision about upward social mobility." Journal of Psychology, 70:81-91.

Toban, Eileen V. 1970.
"Relation betweem future time perspective, age, militancy." Journal of Genetic Psychology, 116:63-68.

Toda, Masanao. 1975.
"Time and the structure of human cognition." In J.T. Fraser and N. Lawrence (eds.). The Study of Time II, pp. 314-324. New York: Springer-Verlag.

Toda, Masanao. 1983.
"Future time perspective and human cognition: an evolutional view." International Journal of Psychology, 18(5):351-365.

Toglia, M.P., and G.A. Kimble. 1976.
"Recall and use of serial position information." Journal of Experimental Psychology: Human Learning and Memory, 2:431-445.

Tolor, Alexander, Gary G. Brannigan, and Vincent M. Murphy. 1970.
"Psychological distance, future time perspective, and internal-external expectancy." Journal of Projective Techniques and Personality Assessment, 34(4):283-294.

Treisman, A.M. 1971.
"Shifting attention between the ears." Quarterly Journal of Experimental Psychology, 23:157-167.

Treisman, Michel. 1963.
"Temporal discrimination and indifference interval: implications for a model of the internal clock." Psychological Monograph, 77, whole number 576.

Treisman, Michel. 1984.
"Contingent aftereffects and situationally coded criteria: discussion paper." Annals of the New York Academy of Sciences, 423:131-141.

Treisman, Michel. 1984.
"Temporal rhythms and cerebral rhythms." Annals of the New York Academy of Sciences, 423:542-565.

Trommsdorff, Gisela. 1983.
"Future orientation and socialization." International Journal of Psychology, 18(5):381-406.

Trommsdorff, Gisela., Helmut Lamm. 1975.
"An analysis of future orientation and some of its social determinants." In J.T. Fraser and N. Lawrence (eds.), The Study of Time II, pp. 343-361. New York: Springer-Verlag.

Trommsdorff, Gisela., Helmut Lamm. 1975.
"Future orientation of institutionalized and

noninstitutionalized delinquents and nondelinquents." European Journal of Social Psychology, 10(3):247-278.

Trommsdorff, Gisela, Helmut Lamm, and Rolf W. Schmidt. 1979. "A longitudinal study of adolescents' future orientation (time perspective)." Journal of Youth and Adolescence, 8(2):131-147.

Tsao, Yao-Chung, E. Wittlieb, B. Miller, and T. Wang. 1983. "Time estimation of a secondary event." Perceptual and Motor Skills, 57:1051-1055.

Tversky, A., and D. Kahneman. 1973. "Availability: a heuristic for judging frequency and probability." Cognitive Psychology, 5:207-232.

Tyldesley, D.A., and H.T.A. Whiting. 1975. "Operational timming." Journal of Human Movement Studies, 1:172-177.

Tysk, Lennart. 1983. "Time estimation by healthy subjects and schizophrenic patients: a methodological study." Perceptual and Motor Skills, 56(3):983-988.

Tysk, Lennart. 1983. "Estimation of time and the subclassification of schizophrenic disorders." Perceptual and Motor Skills, 57(3):911-918.

Tysk, Lennart. 1984. "Time perception and affective disorders." Perceptual and Motor Skills, 58(2):455-464.

Tysk, Lennart. 1984. "A longitudinal study of time estimation in psychotic disorders." Perceptual and Motor Skills, 59(3):779-789.

Tysk, Lennart. 1985. "Longitudinal changes in time estimation in affective disorders: a preliminary study." Perceptual and Motor Skills, 60(1):179-188.

Tzeng, O.J.L., and B. Cotton. 1980. "A study-phase retrieval model of temporal coding."

Journal of Experimental Psychology: Human Learning and Memory, 6:705-716.

Tzeng, O.J.L., A.T. Lee, and C.D. Wetzel. 1979.
"Temporal coding in verbal information processing." Journal of Experimental Psychology: Human Learning and Memory, 5:52-64.

Uchida, Nobuko. 1985.
"Integration and production of causal sequences by preschool children." Japansese Journal of Educational Psychology, 33(2):124-134.

Underwood, B.J. 1977.
Temporal Codes for Memories: Issues and Problems. Hillsdale, NJ: Erlbaum.

Underwood, B.J., and R.M. Malmi. 1978.
"An evaluation of measures used in studying temporal codes for words within a list." Journal of Verbal Learning and Verbal Behavior, 17:279-293.

Underwood, G. 1975.
"Attention and the perception of duration during encoding and retrieval." Perception, 4:291-296.

Usizima, Y. 1935.
"Character and time consciouness." Japanse Journal of Applied Psychology, 3:165-170.

Veach, Tracey L., and John C. Touhey. 1971.
"Personality correlates of accurate time perception." Perceptual and Motor Skills, 33(3):765-766.

Vernon, J., and R. McGill. 1963.
"Time estimation during sensory deprivation." Journal of General Psychology, 69:11-18.

Verstraeten, Damielle. 1980.
"Level of realism in adolescent future time perspective." Human Development, 23(3):177-191.

Vitz, P.S., and T.C. Todd. 1969.
"A coded element model of the perceptual processing of sequential stimuli." Psychological Review, 76:433-449.

Von Franz, M-L. 1981.
"Time and synchronicity in analytic psychology." In J.T. Fraser (ed.), <u>The Voices of Time</u>, 2nd edn., pp. 218-232. Amherst, MA: University of Massachusetts Press.

Von Wright, Johan M., and Riitta Kinnunen. 1976.
"On scaling future time." <u>Perceptual and Motor Skills</u>, 43(3):1235-1241.

Von Wright, Johan M., and Riitta Kinnunen. 1977.
"Scaling of involvement in personal and global future events." <u>Scandinavin Journal of Psychology</u>, 18(4):327-332.

Von Wright, Johan M., and M. Rauste Von Wright. 1977.
"Sex differences in personal and global future time perspectives." <u>Perceptual and Motor Skills</u>, 44(1):30.

Vorberg, Dirk. 1978.
"Problems in the study of response timing. Comments on Reece's (1976) 'A model of temporal tracking'." <u>Acta Psychologica</u>, 42:67-77.

Vorberg, Dirk, and Rolf Hambuch. 1978.
"On the temporal control of rhythmic performance." In J. Requin (ed.), <u>Attention and Performance VII</u>, pp. 535-555. Hillsdale, NJ: Erlbaum.

Vorberg, Dirk, and Rolf Hambuch. 1984.
"Timing of two-handed rhythmic performance." <u>Annals of the New York Academy of Sciences</u>, 423:390-406.

Voyat, Gilbert. 1977.
"Perception and concept of time: a developmental perspective." In Bernard S. Gorman and Alden E. Wessman (eds.), <u>The Personal Experience of Time</u>, pp. 135-160. New York: Plenum.

Vroon, P.A. 1970.
"Effects of presented and processed information on duration experience." <u>Acta Psychologica</u>, 34:115-121.

Vrtunski, Bart P., Marian Patterson, and Gregory O. Hill. 1984.
"Factor analysis of choice reaction time in young and elderly subjects." <u>Perceptual and Motor Skills</u>, 59(2):659-676.

Wagner, John A., and R.J. Katz. 1984.
"The temporal structure of motivation: a within-episodes analysis of priming and provocation." Behavioral and Neural Biology, 41(2):217-222.

Walker, J.T., and A.L. Irion. 1979.
"Two new contingent aftereffects: perceived auditory duration contingent on pitch and on temporal order." Perception and Psychophysics, 26:241-244.

Walker, J.T., A.L. Irion, and D.G. Gordon. 1981.
"Simple and contingent aftereffects of perceived duration in vision and audition." Perception and Psychophysics, 29:475-486.

Wallace, Melvin. 1956.
"Future time perspective in schizophrenia." Journal of Abnormal and Social Psychology, 52:240-245.

Wallace, Melvin., and Albert I. Rabin. 1956.
"Temporal experience." Psychological Bulletin, 57:213-236.

Wallis, Robert. 1967.
"Time: fourth dimension of the mind." Annals of the New York Academy of Sciences, 138:784-797.

Ward, R.A. 1981-82.
"Aging, the use of time, and social change." International Journal of Aging and Human Development, 14(13):177-187.

Warm, J.S., R.M. Stutz, and P.A. Vassolo. 1975.
"International trasnfer in temporal discrimination." Perception and Psychophysics, 18:281-286.

Warner, Donald J., and Richard A. Block. 1984.
"Type A behaviour and temporal judgment." Bulletin of the Psychonomic Society, 22(3):163-166.

Warren, R.M. 1974.
"Auditory temporal discrimination by trained listeners." Cognitive Psychology, 6:237-256.

Warren, R.M., C.J. Obusek, R.M. Farmer, and R.P. Warren. 1969.

"Auditory sequence: confusion of patterns other than speech and music." Science, 144:586-587.

Waterworth, John A. 1985.
"Memory mechanisms and the psychophysical scaling of duration." Perception, 14(1):81-92.

Watts, Fraser N., and Robert Sharrock. 1984.
"Fear and time estimation." Perceptual and Motor Skills, 59(2):597-598.

Wearden, John H., and Charles P. Shimp. 1985.
"Local temporal patterning of operant behavior in humans." Journal of the Experimental Analysis of Behavior, 44(3):315-324.

Weber, A.O. 1933.
"Estimation of time." Psychological Bulletin, 30:233-252.

Webster, Ian W. 1976.
"Aging and the relativity of time." Journal of the American Geriatrics Society, 24(7):314-316.

Weinreb, N., and C.J. Brainerd. 1975.
"A developmental study of Piaget's groupement model of the emergence of speed and time concepts." Child Development, 46:176-185.

Weinstein, Neil D. 1980.
"Unrealistic optimism about future life events." Journal of Personality and Social Psychology, 39:806-820.

Weisberg, P., and C.J. Tragackis. 1967.
"Analysis of DRL behavior in young children." Psychological Reports, 21:709-715.

Weisberg, P. 1970.
"Effects of reinforcement history on timing (DRL) performance in young children." Journal of Experimental Child Psychology, 9:348-362.

Welford, A.T. 1952.
"The 'psychological refractory period' and the timing of high-speed performance -- a review and a theory." British Journal of Psychology, 43:2-19.

Wells, J.E. 1974.
"Strength theory and judgments of recency and frequency." Journal of Verbal Learning and Verbal Behavior, 13:378-392.

Werboff, J. 1962.
"Time judgment as a function of electroencephalographic activity." Experimental Neurology, 6:152-160.

Werner, Carol M., and L.M. Haggard. 1985.
"Temporal qualities of interpersonal relationships." In M.L. Knapp and G.R. Miller (eds.), Handbook of Interpersonal Communication. Beverly Hills, CA: Sage.

Wessman, Alden E. 1973.
"Personality and the subjective experience of time." Journal of Personality Assessment, 37(2):103-114.

Wessman, Alden E., and Bernard S. Gorman. 1977.
"The emergence of human awareness and concepts of time." In Bernard S. Gorman and Alden E. Wessman (eds.), The Personal Experience of Time, pp. 3-55. New York: Plenum.

Wessman, Alden E., Bernard S. Gorman, Gertrude R. Schmeidler, S. Thayer, and E. Mannucci. 1974.
"Personality and linear representatin of temporal location." Journal of Consutling and Clinical Psychology, 42:194.

Westcott, George H., and Lawrence K. Williams. 1976.
"Interpersonal trust and modern attitudes in Peru: work orientation, self-efficacy, and future orientation." International Journal of Contemporary Sociology, 13(1-2):117-137.

Westman, Alida S. 1987.
"Development of time concepts: differentiating clock and calendar from apparent durations." Journal of Genetic Psychology, 148(3):259-270.

Wever, R. 1979.
The Circadian System of Man. New York: Springer-Verlog.

Wever, R. 1982.
"Behavioral aspects of circadian rhythmicity." In F.M. Brown and R.C. Graeber (eds.), Rhythmic Aspects of Behavior, pp. 105-171. Hillsdale, NJ: Erlbaum.

White, C.T. 1963.
"Temporal numerosity and the psychological unit of duration." Psychological Monographs, whole no. 575.

Whyman, A.D., and R.H. Moos. 1967.
"Time perception and anxiety." Perceptual and Motor Skills, 24:567-570.

Wicker, Allan W., and Jeanne C. King. 1988.
"Life cycles of behavior settings." In Joseph E. McGrath (ed.), The Social Psychology of Time: New Perspectives, pp. 182-200. Newbury Park, CA: Sage.

Wilkening, Friedrich. 1981.
"Integrating velocity, time, and distance information: a developmental study." Cognitive Psychology, 13:231-247.

Wilkening, Friedrich. 1982.
"Children's knowledge about time, distance, and velocity interrelations." In William J. Friedman (ed.), The Developmental Psychology of Time, pp. 87-112. New York: Academic.

Wilkinson, R.T. 1982.
"The relationship between body temperature and performance across circadian phase shifts." In F.M. Brown and R.C. Graeber (eds.), Rhythmic Aspects of Behavior, pp. 213-240. Hillsdale, NJ: Erlbaum.

Willette, Thomas L., and George M. Early. 1985.
"Abilities of normal and reading-disabled children to combine the visual and auditory modalities with dimensions of space and time." Perceptual and Motor Skills, 61(3,pt2):1295-1298.

Wilsoncroft, W.E., and Roy S. Griffiths. 1985.
"Time perception and body temperatury: a review." Psychological Documents, 15(2):19.

Wilsoncroft, W.E., J.D. Stone, and F.M. Bagrash. 1978.
"Temporal estimates as a function of difficulty of mental arithmetic." Perceptual and Motor Skills, 46:1311-1317.

Wing, Alan M. 1977.
"Effects of type of movement on the temporal precision of response sequences." British Journal of Mathematical and Statistical Psychology, 30:60-72.

Wing, Alan M. 1978.
"Response timing in handwriting." In G.E. Stelmach (ed.), Information Processing in Motor Control and Learning, pp. 153-172. New York: Academic.

Wing, Alan M. 1979.
"A note on the estimation of the autocovariance function in the analysis of timing of repetitive responses." British Journal of Mathematical and Statistical Psychology, 32:143-145.

Wing, Alan M. 1980.
"The long and short of timing in response sequences." In G.E. Stelmach and J. Requin (eds.), Tutorials in Motor Behavior. Amsterdam, Netherlands: North Holland.

Wing, Alan M. 1982.
"Timing and co-ordination of repetitive bimanual movements." Quarterly Journal of Experimental Psychology, 34A:339-348.

Wing, Alan M., Steven W. Keele, and David I. Margolin. 1984.
"Motor disorder and the timing of repetitive movements." Annals of the New York Academy of Sciences, 423:183-192.

Wing, Alan M., and Alfred B. Kristofferson. 1973.
"Response delays and the timing of discrete motor responses." Perception and Psychophysics, 14:5-12.

Wingfield, A. 1977.
"The perception of alternated speech." Brain and Language, 4:219-230.

Winograd, Eugene, and Robert M. Soloway. 1985.
"Reminding as a basis of temporal judgements." Journal of Experimental Psychology: Learning, Memory, and Cognition, 11(2):262-271.

Witherspoon, Dawn, and Lorraine G. Allan. 1985.
"The effect of a prior presentation on temporal judgments in a perceptual identification task." Memory and Cognition, 13(2):101-111.

Wixen, Bradford D. 1986.
"Subjective time and perceptual skill." Perceptual and Motor Skills, 62(3):873-874.

Wolf, Fredric M., and Mark L. Savickas. 1985. "Time perspective and causal attributions for achievement." Journal of Educational Psychology, 77(4):471-480.

Woodrow, H. 1930. "The reproduction of temporal intervals." Journal of Experimental Psychology, 13:473-499.

Woodrow, H. 1934. "The temporal indifference interval determined by the method of mean error." Journal of Experimental Psychology, 17:167-188.

Woodrow, H. 1935. "The effect of practice upon time-order errors in the comparison of temporal intervals." Psychological Review, 42:127-152.

Woodrow, H. 1951. "Time perception." In S.S. Stevens (ed.), Handbook of Experimental Psychology, pp. 1224-1236. New York: Wiley.

Woodruff, D. 1975. "Relationships among EEG alpha frequency, reaction time, and age: a biofeedback study." Psychophysiology, 12:673-681.

Wyer, Robert S., and Galen V. Bodenhausen. 1985. "Event memory: the effects of processing objectives and time delay on memory for action sequences." Journal of Personality and Social Psychology, 49(2):301-316.

Wyer, Robert S., Edward J. Shoben, Robert W. Fuhrman, and Galen V. Bodenhausen. 1985. "Event memory: the temporal organization of socical action sequences." Journal of Personality and Social Psychology, 49(4):857-877.

Yaker, Henri M. 1972. "The schizophrenic perception of time: a syntactical analysis of time language." In Henri Yaker, Humphry Osmond, and Frances Cheek (eds.), The Future of Time: Man's Temporal Enviroment, pp. 293-309. Garden City, NY: Doubleday.

Yntema, D.B., and F.P. Trask. 1963.
"Recall as a search process." _Journal of Verbal Learning and Verbal Behavior_, 2:65-74.

Yoblick, D.A., and G. Salvendy. 1970.
"Influence of frequency on the estimation of time for auditory, visual, and tactile modalities: the kappa effects." _Journal of Experimental Psychology_, 86:157-164.

Yonge, George D. 1973.
"Time experiences as measures of personality." _Measurement and Evaluation in Guidance_, 5(4):475-482.

Zacks, R.T., L. Hasher, J.W. Alba, H. Sanft, and K.C. Rose. 1984.
"Is temporal order encoded automatically?" _Memory and Cognition_, 12(4):387-394.

Zakay, Dan, and Edi Fallach. 1984.
"Immediate and remote time estimation: a comparison." _Acta Psychologica_, 57(1):69-81.

Zakay Dan, Jacob Lomranz, and Marik Kaziniz. 1984.
"Extraversion-introversion and time perception." _Personality and Individual Differences_, 5(2):237-239.

Zakay, Dan, D. Nitzan, and J. Glicksohn. 1983.
"The influence of task difficulty and external tempo on subjective time estimation." _Perception and Psychophysics_, 34(5):451-456.

Zaleski, Zbigniew. 1987.
"Behaviorial effects of self-set goals for different time ranges." _International Journal of Psychology_, 22(1):17-38.

Zeitlhofer, Josef, B. Saletu, J. Stary, and Ramazamali Ahmadi. 1984.
"Cerebral function in hyperthyroid patients: psychopathology, psychometric variables, central arousal and time perception before and after thyrostatic therapy." _Neuropsychobiology_, 11(2):89-93.

Zepelin, Harold. 1986.
"REM sleep and the timing of self-awakenings." _Bulletin of Psychonomic Society_, 24(4):254-256.

Zern, D. 1967.
"The influence of certain developmental factors in fostering the ability to differentiate the passage of time." _Journal of Social Psychology_, 72:9-17.

Zern, D. 1970.
"The influence of certain child-rearing factors upon the development of a structured and salient sense of time." _Genetic Psychology Monographs_, 81, No. 2, pp. 197-254.

Zhu, Qi. 1985.
"The knowledge and experience regulation in the process of olfactory hallucination and the constancy of time perception." _Information on Psychological Sciences_, 6:40-44.

Ziller, Robert C. 1977.
"Group dialetics: the dynamics of groups over time." _Human Development_, 20(5):293-308.

Zimmerman, J., and B.J. Underwood. 1968.
"Ordinal position knowledge within and across lists as a function of instructions in free-recall learning." _Journal of General Psychology_, 79:301-307.

Zurcher, Jr., Louis A., Joe E. Willis, and Fredrick F. Ikard. 1967.
"Dogmatism, future orientation, and perception of time." _Journal of Social Psychology_, 73(2):205-209.

19

Sociology

Adler, P. 1981.
Momentum: A Theory of Social Action. Beverly Hills, CA: Sage.

Agblemagnon, F. N'Sougan. 1979.
"Time in rural societies and rapidly changing societies." In Frank Greenaway (ed.), Time and the Sciences, pp. 163-173. Paris, France: Unesco Press.

Aguessy, Honorat. 1977.
"Sociological interpretations of time and pathology of time in developing countries." Time and the Philosophies, pp. 94-105. Paris, France: Unesco Press.

Airaksinen, Timo. 1985.
"Social time and place." Man and World, 18:99-105.

Albert, S., and S. Kessler. 1976.
"Processes for ending social encounters: the conceptual archeology of a temporal place." Journal of the Theory of Social Behavior, 6 (October):147-70.

Albert, Stuart, and William Jones. 1977.
"The temporal transition from being together to being alone: the significance and structure of children's bedtime stories." In Bernard S. Gorman and Alden E. Wessman (eds.), The Personal Experience of Time, pp. 111-132. New York: Plenum.

Allen, Francis R. 1971.
Socio-Cultural Dynamics: An Introduction to Social Change. New York: Macmillan.

Armstrong, David. 1985.
"Space and time in British general practice." Social Science and Medicine, 20(7):659-666.

Bartky, Ian R. 1984.
"A comment on 'The standardization of time' by Zerubavel." American Journal of Sociology, 89(6):1420-1425.

Beach, Betty A. 1987.
"Time use in rural home-working families." Family Relations, 36(4):412-416.

Beach, Wayne A. 1985.
"Temporal density in the courtroom interaction: constraints on the recovery of past events in legal discourse." Communication Monographs, 52(1):1-18.

Becerra, Rosina M., and D. Shaw. 1984.
The Hispanic Elderly: A Research Reference Guide. Washington, DC: University Press of America.

Belloni, Maria Carmen. 1986.
"Social time dimensions as indicators of class distinction in Italy." International Social Science Journal, 107:65-76.

Bengston, V. 1983.
Time, aging and the continuity of social structure." Journal of Social Issues, 39(4):45-72.

Bennis, Warren G., and Philip E. Slater. 1968.
The Temporary Society. New York: Harper and Row.

Bevans, G. de. 1953.
How Working Men Spend Their Spare Time. New York: Columbia University Press.

Blakelock, E.H. 1961.
"A Durkheimian approach to some temporal problems of leisure." Social Problems, 9.

Blyton, F. 1985.
Changes in Working Time: An International Review. London, UK: Croom Helm.

Blyton, F. 1985.
"Workplace democracy, unemployment and the reduction of working time." Economic and Industrial Democracy, 6(1):113-20.

Bock, Philip K. 1964.
"Social structure and language structure." Southwestern Journal of Anthropology, 20:393-403.

Bock, Philip K. 1966.
"Social time and institutional conflict." Human Organization, 25 (Summer):96-102.

Bortner, Raymond W., et al. 1974.
"A cross-cultural study of the effects of children on parental assessment of past, present and future." Journal of Marriage and the Family, 36:98-103.

Bosserman, Philip. 1975.
"Some theoretical dimensions of the study of social time." Society and Leisure, 7(1):21-44.

Brim, O.G., and R.A. Forer. 1956.
"A note on the relation of values and social structure to life planning," Sociometry, 19:54-60.

Bronfenbrenner, Urie. 1958.
"Social class through time and space." In Eleanor E. Maccoby et al. (eds.), Readings of Social Psychology, pp. 400-425. New York: Holt, Rinehart and Winston.

Bull, C. Neil. 1978.
"Chronology: the field of social time." Journal of Leisure Research, 10:294.

Calkins, K. 1970.
"Time: perspectives, marking and styles of usuage." Social Problems, 17:487-501.

Chiswick, C.U. 1982.
"Value of a housewife's time." Journal of Human Resources, 17 (Summer):413-25.

Colsen, F.H. 1926.
The Week. Cambridge, UK: Cambridge University Press.

Cottle, Thomas J. 1969.
"Future orientations and avoidance: speculations on the time of achievement and social roles." Sociological Quarterly, 10(4):419-437.

Cottrell, W.F. 1939.
"On time and railroader." American Sociological Review, 4:190-198.

Couch, Carl J., et al. 1975.
"Time and social life." In Carl J. Couch (ed.), Constructing Social Life, pp. 120-138. Champaign, IL: Stipes.

Couch, Carl J., 1982.
"Temporality and paradigms of thought." Studies in Symbolic Interaction, 4:1-33.

Couch, Carl, J., and Robert A. Hintz, Jr. (eds.). 1975.
"Time, intention and social behavior." In Carl J. Couch and Robert A. Hintz, Jr. (eds.), Constructing Social Life, pp. 27-46. Champaign, IL: Stipes.

Couch, Carl J., Marilyn G. Leichty, Leigh Anderson, and Robert A. Hintz, Jr. 1975.
"Time and social life." In Carl Couch and Robert A. Hintz, Jr. (eds.), Constructing Social Life, pp. 120-138. Champaign, IL: Stipes.

Crandall, Rick. 1976.
"Future directions for the psychology of leisure." Society and Leisure, 8(3):105-09.

Czarnocki, Dan. 1978.
"Macrotime, midi-time and micro-time: a set of decompositional techniques for making historical sense out of longitudinal data." Canadian Journal of Sociology, 3(1):21-39.

Denzin, Norman K. 1987.
"Under the influence of time: reading the interactional text." Sociological Quarterly, 28(3):327-341.

D'Epinay, Christian Lalive. 1986.
"Time, space and socio-cultural identity: the ethos of the proletariat, small owners and peasantry in an aged population." International Social Science Journal, 107:89-104.

Deva, Indra. 1984.
"A sociological perspective of time and creativity." Impact of Science on Society, 2-3(134-135):197-202.

Dumazedier, Joffre. 1960.
"Sociological aspects of leisure." International Social Science Journal, 12(4):517-28.

Dumazedier, Joffre. 1967.
Toward a Sociology of Leisure. New York: Free Press.

Elchardus, Mark. 1988.
"The rediscovery of chronos: the new role of time in sociological theory." International Sociology, 3(1):35-59.

Engel, David M. 1987.
"Law, time, and community." Law and Society Review, 21(4):605-637.

Engel-Frisch, Gladys. 1943.
"Some neglected temporal aspects of human ecology." Social Forces, 22:43-47.

Felson, Marcus. 1980.
"Human chronography." Sociology and Social Research, 65(1):1-9.

Filipcova, Blanka, and Jindrich Filipec. 1986.
"Society and concepts of time." International Social Science Journal, 107:19-32.

Firey, Walter. 1963.
"Conditions for the realization of values remote in time." In Edward A. Tiryakian (ed.), Sociological Theory, Values, and Sociological Change, pp. 147-159. New York: Free Press.

Folkard, S., and T. Monk. 1985.
Hours of Work: Temporal Factors in Work Scheduling. New York: Wiley.

Fortes, Meyer. 1970.
Time and Social Structure and Other Essays. Atlantic Highlands, NJ: Humanities.

Frankenberg, Ronald. 1988.
"Your time or mine? An anthropological view of the tragic temporal contradictions of biomedical practice." International Journal of Health Services, 18(1):11-34.

Fraser, J.T. 1978.
"The individual and society." In J.T. Fraser, N. Lawrence, and D. Park (eds.), The Study of Time III, pp. 419-442. New York: Springer-Verlag.

Gabennesch, Howard. 1988.
"When promises fail: a theory of temporal fluctuations in suicide." Social Forces, 67(1):129-145.

Gelles, Richard J., and R.R. Faulkner. 1978.
"Time and television news work: task temporalization in the assembly of unscheduled events." Sociological Quarterly, 19(1):89-102.

Germain, Carel B. 1983.
"Time, social change, and social work." Social Work in Health Care, 9(2):15-23.

Gershuny, Jonathan I. 1986.
"Time use, technology, and the future of work." Journal of the Market Research Society, 28(4):335-354.

Giddens, Anthony. 1981.
"Time and space in social theory: critical remarks upon functionalism." Current Perspectives in Social Theory, 2:3-13.

Gihring, T.A. 1983.
"Leisure-time activities in an urban Nigerian setting: attitudes and experience." Journal of Leisure Research, 15(2):108-24.

Gioscia, Victor. 1972.
"On social time." In Henri Yaker, Humphrey Osmond, and Frances Cheek (eds.), The Future fo Time: Man's Temporal Environment, pp. 73-141. Garden City, NY: Doubleday.

Glasser, Richard. 1972.
Time in French Life and Thought. Manchester, UK: Manchester University Press.

Glassner, Barry. 1982.
"An essay on interative social time." Sociological Review, 30(N.S.):668-681.

Goody, Jack. 1968.
"Time: social organization." In David L. Sills (ed.), International Encyclopedia of the Social Sciences, Vol. 16, pp. 30-42. New York: Macmillan & Free Press.

Gregory, Jr., Stanford W. 1983.
"A quantitative analysis of temporal symmetry in microsocial relations." American Sociological Review, 48:129-135.

Gross, David. 1985.
"Temporality and the modern state." Theory and Society, 14(1):53-82.

Grudin, Robert. 1988.
Time and the Art of Living. New York: Ticknor & Fields.

Gurvitch, Georges. 1963.
"Social structure and the multiplicity of times." In Edward A. Tiryakian (ed.), Sociological Theory, Values, and Sociocultural Change, pp. 171-184. New York: Free Press.

Gurvitch, Georges. 1964.
The Spectrum of Social Time. Dordrecht, Netherlands: D. Reidel.

Hantrais, L. 1985.
"Leisure lifestyles and the synchronisation of family schedules." World Leisure and Recreation, 27(2):18-24.

Hantrais, L., Peter A. Clark, and Nicole Samuel. 1984.
"Time-space dimensions of work, family and leisure in France and Great Britain." Leisure Studies, 3(3):301-317.

Hardesty, Monica J. 1982.
"Ethnomethodlogy and symbolic interactionism: a critical

comparison of temporal orientations." Symbolic Interaction, 5(1):127-137.

Harvey, A., et al. 1984.
Time Budget Research. Berlin, FRG: Campus Verlag.

Hayden, F. Gregory. 1980.
"A critical analysis of time stream discounting for social program evaluation." Social Science Journal, 17(1):21-40.

Hearn, Francis. 1980.
"Past, play and future." Free Inquiry in Creative Sociology, 8(1):94-99.

Heirich, Max. 1964.
"The use of time in the study of social change." American Sociological Review, 29:386-398.

Hereven, Tamara K. 1977.
"Family time and historical time." Daedalus, 106, No. 2., pp. 57-70.

Hereven, Tamara K. 1982.
Family Time and Industrial Time: The Relationship Between the Family and Work in a New England Industrial Community. Cambridge, UK: Cambridge University Press.

Horton, John. 1967.
"Time and cool people." Trans-action, 4(5):5-12.

Hulett, Jr., J.E. 1944.
"The person's time perspective and the social role." Social Forces, 23:155-159.

Julkunen, Raija. 1977.
"A contribution to the categories of social time and the economy of time." Acta Sociologica, 20:5-24.

Kaplan, Max, and Philip Bosserman (ed.). 1971.
Technology, Human Values, and Leisure. Nashville, TN: Abingdon Press.

Katovich, Michael A. 1986.
"Temporal stages of situated activity and identity activation." Studies in Symbolic Interaction, 2(B):329-352.

Katovich, Michael A. 1987.
"Durkheim's macrofoundations of time: an assessment and critique." Sociological Quarterly, 28(3):367-385.

Kolaja, Jiri. 1969.
Social System and Time Space. Pittsburgh, PA: Duquesne University Press.

La-Rossa, Ralph. 1983.
"The transition to parenthood and the social reality of time." Journal of Marriage and Family, 43(3):579-589.

Lauer, Robert H. 1973.
"Temporality and social change: the case of 19th century China and Japan." Sociological Quarterly, 14(4):451-464.

Lauer, Robert H. 1981.
Temporal Man: The Meaning and Uses of Social Time. New York: Praeger.

Le Goff, Jacques. 1970.
"Church time and merchant time in the middle ages." Social Science Information, 9:151-168.

Le Goff, Jacques. 1980.
Time, Work and Culture in the Middle Ages. Chicago, IL: University of Chicago Press.

Levine, Robert V. 1985.
"Social time and the pace of life on four continents." Social Science Newsletter, 70(2):71-76.

Lewis, J. David, and Andrew J. Weigert. 1981.
"The structures and meanings of social time." Social Forces, 60:432-462.

Little, V. 1983.
"An overview of research using the time budget methodology," Aging and Society, 4(1).

Luhmann, Niklas. 1976.
"The future cannot begin: temporal structures in modern society." Social Research, 43(1):130-152.

Luscher, Kurt K. 1974.
"Time: a much neglected dimension in social theory and research." Sociological Analysis and Theory, 4:101-116.

Maines, David R. 1987.
"The significance of temporality for the development of sociological theory. Sociological Quarterly, 28(3):303-311.

Maines, David R., Noreen M. Sugrue, and Michael A. Katovich. 1983.
"The sociological import of G.H. Mead's theory of the past." American Sociological Review, 48:161-73.

Mayr, Albert. 1989.
"Social time in experimental music and art." In J.T. Fraser (ed.), Time and Mind: The Study of Time VI. Madison, CT: International Universities Press.

McGrath, Joseph E. 1986.
"Continuity and change: time method and the study of social issues." Journal of Social Issues, 42(4):5-19.

McGukin, J. Drew, and William J. Seiler. 1987.
"The temporal organization of classrooms as an interactional accomplishment." Journal of Thought, 22(4):106-113.

Meier, R.L. 1959.
"Human time allocation: a basis for social accounts." Journal of the American Institute of Planners, 15:27-33.

Melbin, Murray. 1978.
"Night as frontier." American Sociological Review, 43:3-22.

Melbin, Murray. 1978.
"City rhythms." In J.T. Fraser, et al., (eds.), The Study of Time III, pp. 444-465. New York: Springer-Verlag.

Melbin, Murray. 1984.
"Issues in using 24 hours of the day." Research in Social Problems and Public Policy, 3:151-168.

Melbin, Murray. 1987.
Night as a Frontier: Colonizing the World After Dark. New York: Free Press.

Meyerson, Fred W. 1969.
"The sociology of leisure in the United States, 1945-1965." Journal of Leisure Research, 1 (Winter):53-69.

Michelson, William. 1971.
"Some like it hot: social participation and environmental use as functions of the season." American Journal of Sociology, 76:1072-1083.

Michon, Francois. 1987.
"Time and flexibility: working time in the debate on flexibility." Labour and Society, 12(1):154-174.

Mol, Hans. 1982.
"Time and transcendence in a dialectical sociology of religion." Sociological Analysis, 42(4):317-324.

Moore, Wilbert E. 1963.
Man, Time, and Society. New York: Wiley.

Moore, Wilbert E. 1963.
"The temporal structure of organizations." In Edward A. Tiryakian (ed.), Sociological Theory, Values, and Sociocultural Change, pp. 161-169. New York: Free Press.

Moore, Wilbert E. 1967.
"Forecasting the future." In W.E. Moore, (ed.), Order and Change: Essays in Comparative Sociology. New York: Wiley.

Mukerjee, Radhakamal. 1943.
"Time, technics, and society." Sociology and Social Research, 27:261.

Musgrove, Frank, and Roger Middleton. 1981.
"Rites of passage and the meaning of age in three contrasted social groups: professional footballers, teachers, and Methodist ministers." British Journal of Sociology, 32:39-55.

Nadel, George H. 1968.
"Periodization." In David L. Sills (ed.), International Encyclopedia of the Social Sciences, Vol. 11, pp. 581-585.

Nowotny, Helga. 1975.
"Time structuring and time measurement: on the interrelation between timekeepers and social time." In J.T. Fraser and N. Lawrence (eds.), The Study of Time II, pp 325-342. New York: Springer-Verlag.

Nowotny, Helga. 1989.
"Mind, technologies, and collective time consciousness: from the future to an extended present." In J.T. Fraser (ed.), Time and Mind: The Study of Time VI. Madison, CT: International Universities Press.

O'Rand, Angela, and Robert A. Ellis. 1974.
"Social class and social time perspective." Social Forces, 53(1):53-62.

Patrushev, V.D. 1986.
"Past and future changes in Soviet workers' time-budget." International Social Science Journal, 107:77-88.

Persson, M. 1981.
"Time-perspectives amongst criminals." Acta Sociologica, 24(3):149-166.

Pratt, Lois. 1981.
"Business temporal norms and bereavement behavior." American Sociological Review, 46(3):317-333.

Pronovost, Gilles. 1986.
"Time in a sociological and historical perspective." International Social Science Journal, 107:5-18.

Rezsohazy, Rudolf. 1972.
"The concept of social time: its role in development." International Social Science Journal, 24:26-36.

Rezsohazy, Rudolf. 1986.
"Recent social developments and changes in attitudes to time." International Social Science Journal, 107:33-48.

Rigby, P. 1983.
"Time, historical consciousness and the Masai." Comparative Studies in Sociology and History, 25:428-456.

Rindfuss, Ronald R., S. Philip Morgan, and C. Gray Swicegood. 1984.
"The transition to motherhood: the intersection of

structural and temporal dimensions." American Sociological Review, 49(3):359-372.

Robinson, John P. 1977.
How Americans Use Time: A Social-Psychological Analysis of Everyday Behavior. New York: Praeger.

Robinson, John P. 1988.
"Time-diary evidence about the social psychology of everyday life." In Joseph E. McGrath (ed.), The Social Psychology of Time: New Perspectives, pp. 134-148. Newbury Park, CA: Sage.

Robinson, John P., and Philip E. Converse. 1972.
"Social change reflected in the use of time." In Augus Campbell and Philip E. Converse (eds.), The Human Meaning of Social Change, pp. 17-86. New York: Russell Sage Foundation.

Rosengren, William R., and Spencer De Vault. 1963.
"The sociology of time and space in an obstetrical hospital." In Eliot Freidson (ed.), The Hospital in Modern Society, pp. 266-292. New York: Free Press.

Roth,, Julius A.
Timetables: Structuring the Passage of Time in Hospital Treatment and Other Careers. New York: Bobbs-Merrill.

Rustom, C. 1961.
"The later years of life and the use of time among the Burmane." In R.W. Kleemeier (ed.), Aging and Leisure, pp. 100-103. New York: Oxford University Press.

Samuel, Nicole. 1986.
"Free time in France: a historical and sociological survey." International Social Science Journal, 107:49-63.

Schlesinger, Philip. 1977.
"Newsmen and their time-machine." British Journal of Sociology, 28:336-350.

Schmidt-Glintzer, Helwig. 1986.
"The nature of time schedules in traditional China." In J.T. Fraser, et. al. (eds.), Time, Science, and Society in China and the West: The Study of Time V. Amherst, MA: University of Massachusetts Press.

Schwartz, Barry. 1974.
"Waiting, exchange, and power: the distribution of time in social systems." American Journal of Sociology, 79:841-871.

Schwartz, Barry. 1975.
Queueing and Waiting: Studies in the Social Organization of Access and Delay. Chicago, IL: University of Chicago Press.

Schwartz, Barry. 1978.
"Queues, priorities, and social process." Social Psychology, 41:3-12.

Schwartz, Barry. 1978.
"The social ecology of time barriers." Social Forces, 56(4):1203-1220.

Schwartz, Barry. 1978.
"Time, patience, and black people: a study of temporal access to medical care." Sociological Focus, 11:11-20.

Sharron, Avery. 1981.
"Frame paralysis: when time stands still." Social Research, 48:500-520.

Sharron, Avery. 1982.
"Dimensions of time." Studies in Symbolic Interaction, 4:63-89.

Sirianni, Carmen J. 1987.
"Economies of time in social theory: three approaches compared." Current Perspectives in Social Theory, 8:161-195.

Snow, Robert P., and D. Brissett. 1986.
"Pauses: explorations in social rhythm." Symbolic Interaction, 9(1):1-18.

Sorokin, Pitirim A. 1939.
Time-budgets of Human Behavior. Cambridge, MA: Harvard University Press.

Sorokin, Pitirim A. 1943.
Sociocultural Causality, Space and Time. Durham, NC: Duke University Press.

Sorokin, Pitirim, and Robert K. Merton. 1937.
"Social time: a methodological and functional analysis." _American Journal of Sociology_, 42:615-629.

Spangler, M. 1981.
"Time proverbs and social change in Belgrade, Yugoslavia." In J.T. Fraser, et al. (eds.), _The Study of Time IV_, pp. 223-233. New York: Springer-Verlag.

Telles, Joel L. 1980.
"Time, rank, and social control." _Sociological Inquiry_, 50(2):171-183.

Thompson, E.P. 1967.
"Time, work-discipline, and industrial capitalism." _Past and Present_, 38:56-97.

Thornton, Robert J. 1980.
Space, Time, and Culture among the Iraqw of Tanzania. New York: Academic.

Thornton, Robert J. 1989.
"Time scales and social thought." In J.T. Fraser (ed.), _Time and Mind: The Study of Time VI_. Madison, CT: International Universities Press.

Tillman, Mary Katherine. 1970.
"Temporality and role-taking in G.H. Mead." _Social Research_, 37(4):533-546.

Tripi, Frank. 1985.
"Client power and Italian bureaucracy: time delay and its cost to clients." _Social Science Journal_, 22(2):95-104.

Wallis, Robert. 1970.
"Chronopolitics: the impact time perspectives on the dynamics of change." _Social Forces_, 49(1):102-108.

Webster, Hutton. 1916.
Rest Days. New York: Macmillan.

Westcott, George H., and Lawrence K. Williams. 1976.
"Interpersonal trust and modern attitudes in Peru: work orientation, self-efficacy, and future orientation." _International Journal of Contemporary Sociology_, 13(1-2):117-137.

Winthrop, Henry. 1968.
"The sociologist and the study of the future." American Sociologist. 3(2):136-145.

Wright, Lawrence. 1969.
Clockwork Man: The Story of Time, Its Origins, Its Uses, Its Tyranny. New York: Horizon Press.

Young, Michael. 1988.
The Metronomic Society: Cyclical and Linear Time. Cambridge, MA: Harvard University Press.

Young, Michael, and Tom Schuller (eds.). 1988.
The Rhythms of Society. New York: Routledge Chapman and Hall.

Young, Michael, and Jiri Zeman. 1971.
"Cycles in social behavior." Nature, 229:91-95.

Zerubavel, Eviatar. 1976.
"Timetables and scheduling: on the social organization of time." Sociological Inquiry, 46(2):87-94.

Zerubavel, Eviatar. 1977.
"The French republican calender: a case study in the sociology of time." American Sociological Review, 42:868-877.

Zerubavel, Eviatar. 1978.
"Ending social Encounters." Journal of Experimental Social Psychology, 14:541-53.

Zerubavel, Eviatar. 1979.
"Private time and public time: the temporal structure of social accessibility and professional commitments." Social Forces, 58(1):38-58.

Zerubavel, Eviatar. 1979.
Patterns of Time in Hospital Life: A Sociological Perspective. Chicago, IL: University of Chicago Press.

Zerubavel, Eviatar. 1980.
"The Benedictine ethic and the modern spirit of scheduling: on schedules and social life." Sociological Inquiry, 50:157-169.

Zerubavel, Eviatar. 1981.
Hidden Rhythms: Schedules and Calendars in Social Life. Chicago, IL: University of Chicago Press.

Zerubavel, Eviatar. 1982.
"The standardization of time: a sociohistorical perspective." American Journal of Sociology, 88(1):1-23.

Zerubavel, Eviatar. 1987.
"The language of time: toward a semiotics of temporality." Sociological Quarterly, 28(3):343-356.

20
Speech and Communication

Allan, G.D. 1975.
"Speech rhythm." Journal of Phonetics, 3:75-86.

Barry, W.J. 1984.
"Segment or syllable? A reaction-time investigation of phonetic processing." Language and Speech, 27(1):1-15.

Beach, Wayne A. 1985.
"Temporal density in the courtroom interaction: constraints on the recovery of past events in legal discourse." Communication Monographs, 52(1):1-18.

Beattie, Geoffrey W. 1978.
Sequential temporal patterns of speech and gaze in dialogue." Semiotica, 23:29-52.

Beattie, Geoffrey W. 1979.
"The modifiability of the temporal structure of spontaneous speech." In Aron W. Siegman and Stanley Feldstein (eds.), Of Speech and Time: Temporal Speech Patterns in Interpersonal Contexts, pp. 115-148. Hillsdale, NJ: Erlbaum.

Benguerel, Andre Pierre, and Janet D'Arcy. 1986.
"Time-warping and the perception of rhythm in speech." Journal of Phonetics, 14(2):231-246.

Bernstein, N. 1967.
The Coordination and Regulation of Movements. New York: Pergamon.

Braud, Willian G., and Stephen H. Holborn. 1966.
"Temporal content effects with two judgemental languages." Psychonomic Science, 6:151-152.

Bruneau, Thomas. J. 1973.
"Communicative silences: forms and functions." Journal of Communication, 23(1):17-46.

Bruneau, Thomas, J. 1974.
"Time and nonverbal communication." Journal of Popular Culture, 8:658-666.

Bruneau, Thomas, J. 1977.
"Chronemics: the study of time in human interaction (with a glossary of chronemic terminology)." Communication: Journal of the Communication Association of the Pacific, University of Hawaii, 6:1-30.

Bruneau, Thomas, J. 1979.
"Chronemics: time and organizational communication." Communication: Journal of the Communication Association of the Pacific, University of Hawaii, 7:81-97.

Bruneau, Thomas, J. 1979.
"The time dimension in intercultural communication." In D. Nimmo (ed.), Communication Yearbook 3. New Brunswick, NJ: Transition Books.

Bruneau, Thomas. J. 1980.
"Chronemics and the verbal-nonverbal interface." In M.R. Key (ed.), The Relationship of Verbal and Nonverbal Communication, pp. 101-117. Hague, Netherlands: Mouton.

Bruneau, Thomas, J. 1980.
"Theoretical perspectives on the temporal bases of intercultural communicaiton." Communication: Journal of the Communication Association of the Pacific, University of Hawaii, 9(1):77-115.

Bruneau, Thomas, J. 1983.
"Personal time: hello-goodbye-in-between." Communication: Journal of the Communication Association of the Pacific, University of Hawaii, 12(1):39-49.

Clark, E.V. 1971.
"In the acquisition of meaning of _before_ and _after_." _Journal of Verbal Learning and Verbal Behavior_, 10:266-275.

Cohen, Jonathan. 1950-51.
"Tense usage and propositions." _Analysis_, 11.

Coker, P.L. 1978.
"Syntactic and semantic factors in the acquisition of before and after." _Journal of Child Language_, 5:261-277.

Condon, William S. 1978.
"An analysis of behavioral organization." _Sign Language Studies_, 13.

Davis, Martha (ed.). 1982.
Interaction Rhythms: Perodicity in Communication Behavior. New York: Human Sciences Press.

Fay, W.H. 1966.
Temporal Sequence in the Perception of Speech. Hague, Netherlands: Mouton.

Feldstein, Stanley, Luciano Alberti, and Mohammed Bendebba. 1979.
"Self-attributed personality characteristics and the pacing of conversational interaction." In Aron W. Siegman and Stanley Feldstein (eds.), _Of Speech and Time: Temporal Speech Patterns in Interpersonal Contexts_, pp. 73-87. Hillsdale, NJ: Erlbaum.

Feldstein, Stanley, and Joan Welkowitz. 1978.
"A chronography of conversation: in defense of an objective approach." In Aron W. Siegman and Stanley Feldstein (eds.), _Non-verbal Behavior and Communication_. Hillsdale, NJ: Erlbaum.

Fowler, C. 1980.
"Coarticulation and theories of extrinsic timing." _Journal of Phonetics_, 8:113-133.

Frentz, Thomas S. 1985.
"Rhetorical conversation, time, and moral action." _Quarterly Journal of Speech_, 71(1):1-18.

Goldman-Eisler, F. 1967.
"Sequential temporal patterns and cognitive processes in speech." Language and Speech, 10:122-132.

Hall, Joseph W., and Elizabeth J. Wood. 1984.
"Stimulus duration and frequency discrimination for normal-hearing and hearing-impaired subjects." Journal of Speech and Hearing Research, 27(2):252-256.

Hayes, Donald P., and Loren Cobb. 1979.
"Ultradian biorhythms in social interaction." In Aron W. Siegman and Stanley Feldstein (eds.), Of Speech and Time: Temporal Speech Patterns in Interpersonal Contexts, pp. 57-70. Hillsdale, NJ: Erlbaum.

Henderson, A., F. Goldman-Eisler, and A. Skarbek. 1966.
"Sequential temporal patterns in spontaneous speech." Language and Speech, 9:207-216.

Huggins, A. 1978.
"Speech timing and intelligibility." In J. Requin (ed.), Attention and Performance VII, pp. 279-297. Hillsdale, NJ: Erlbaum.

Humes, Larry E., and Jayne B. Ahlstrom. 1984.
"Relation between reaction time and loudness." Journal of Speech and Hearing Research, 27(2):306-310.

Idson, W.L., and Dominic W. Massaro. 1980.
"The role of perceived duration in the identification of vowels." Journal of Phonetics, 8:407-425.

Jaffe, J., and Stanley Feldstein. 1970.
Rhythms of Dialogue. New York: Academic.

Johnson, Charles C., Harry Hollien, and James W. Hicks. 1984.
"Speaker identification utilizing selected temporal speech features." Journal of Phonetics, 12(4):319-326.

Kamp, Hans. 1979.
"Instants, events and temporal discourse." In R. Bauerle, E. Ugli, and A. von Stechow (eds.), Semantics From Different Points of View, pp. 376-417. Berlin, FRG: Springer-Verlag.

Kent, R., and F. Minifie. 1977.
"Coarticulation in recent speech production models." Journal of Phonetics, 5:115-133.

Klatt, D.H. 1975.
"Vowel length is syntactically determined." Journal of Phonetics, 3:129-140.

Klatt, D.H. 1976.
"Linguistic uses of segmental duration in English: acoustic and perceptual evidence." Journal of the Acoustical Society of America, 59:1208-1221.

Kowal, S., D.C. O'Connell, and E.J. Sabin. 1975.
"Development of temporal patterning and vocal hesitations in spontaneous narratives." Journal of Psycholinguistic Research, 4:195-207.

Lehiste, I. 1970.
Suprasegmentals. Cambridge, MA: MIT Press.

Lehiste, I. 1977.
"Isochrony reconsidered." Journal of Phonetics, 5:253-265.

Loots, M. 1980.
Metrical Myths. Hague, Netherlands: Martinus Nijhoff.

Marcus, S.M. 1984.
"Recognizing speech: on the mapping from sound to word." In H. Bouma and D.N.G. Bouwhuis (eds.), Attention and Performance X, pp. 151-163. Hillsdale, NJ: Erlbaum.

Marslen-Wilson, W.D. 1980.
"Speech understanding as a psychological process." In J.D. Simon (ed.), Spoken Language Generation and Recognition, pp. 39-67. Dordrecht, Netherlands: Reidel.

Martin, J.G. 1972.
"Rhythmic (hierarchical) versus serial structure in speech and other behavior." Psychological Review, 79:487-509.

Nooteboom, Sieb G. 1985.
"A functional view of prosodic timing in speech." In John A. Michon and Janet L. Jackson (eds.), Time, Mind, and Behavior, pp. 212-252. New York: Springer-Verlag.

Peterson, G.E., and I. Lehiste. 1960.
"Duration of syllable nuclei in English." Journal of the Acoustical Society of America, 32:693-703.

Schwartz, J., and J. Jaffe. 1968.
"Markovian prediction of sequential temporal patterns in spontaneous speech." Language and Speech, 11:27-30.

Siegman, Aron Wolfe, and Stanley Feldstein (eds.). 1979.
Of Speech and Time: Temporal Speech Rhythms in Interpersonal Contexts. Hillsdale, NJ: Erlbaum.

Sussman, H., and J. Westbury. 1981.
"The effects of antagonistic gestures on temporal and amplitude parameters of anticipatory labial coarticulation." Journal of Speech and Hearing Research, 24:16-24.

Swingle, Paul G., and Gordon F. Hope. 1987.
"Interpersonal influence on temporal measures of vocalization in interview situations." Language and Speech, 30(2):159-167.

Warner, R.M. 1979.
"Periodic rhythms in conversational speech." Language and Speech, 22:381-396.

Welkowitz, Joan, Ronald N. Bond, and Stanley Feldstein. 1984.
"Conversational time patterns of Hawaiian children as a function of ethnicity and gender." Language and Speech, 27(2):173-191.

Welkowitz, Joan, Ronald N. Bond, and Stanley Feldstein. 1984.
"Conversational time patterns of Japanese-American adults and children in same and mixed-gender dyads." Journal of Language and Social Psychology, 3(2):127-138.

Wingfield, A. 1977.
"The perception of alternated speech." Brain and Language, 4:219-230.

21

Miscellaneous

Abraham, Josefina Mena. 1979.
"The unifying time of environmental transformation." In Frank Greenaway (ed.), Time and the Sciences, pp. 143-162. Paris, France: Unesco Press.

Agren, Hans. 1986.
"Chinese traditional medicine: temporal order and synchronous events." In J.T. Fraser, N. Lawrence, and F.C. Haber (eds.), Time, Science, and Society in China and the West: The Study of Time V, pp. 211-2118. Amherst, MA: University of Massachusetts Press.

Barry, W.B.N. 1968.
Growth of a Prehistoric Time Scale, Based on Organic Evolution. San Francisco, CA: Freeman.

Benthem, Johan Van. 1985.
"Semantics of time." In John A. Michon and Janet L. Jackson (eds.), Time, Mind, and Behavior, pp. 266-278. New York: Springer-Verlag.

Blum, Harold F. 1967.
"Humanity in the perspective of time." Annals of the New York Academy of Sciences, 138:489-503.

Calder, Nigel. 1983.
Timescale: An Atlas of the Fourth Dimension. New York: Viking Press.

Campbell, Jeremy. 1986.
Winston Churchill's Afternoon Nap: A Wide-Awake Inquiry Into the Human Nature of Time. New York: Simon and Schuster.

Campbell, John. 1954.
"Functional organization of the central nervous systems with respect to organziation in time." Neurology, 4: 295-300.

Chalendar, Jacques de. 1976.
Lifelong Allocation of Time: Report. Paris, France: OECD.

Chari, C.T.K. 1963.
"Information, cosmology and time." Dialectica, 17:369-380.

Christensen, Harold T., and George R. Carpenter. 1962.
"Timing patterns in the development of sexual intimacy: an attitudinal report on three modern western societies." Marriage and Family Living, 24:30-35.

Clausen, Johs. 1949.
"Time judgement." In Fred A. Mettler (ed.), Selective Partial Ablation of the Frontal Cortex, pp. 254-256. New York: Hoeber.

Coheen, Jack L. 1950.
"Disturbances in time discrimination in organic brain disease." Journal of Nervous and Mental Diseases, 112:121-129.

Couch, Carl J. 1984.
Constructing Civilizations. Greenwich, CT: JAI Press.

Davies, Paul C.W. 1983.
"Inflation and time asymmetry in the universe." Nature, 310:398-400.

De Long, Alton J. 1985.
"Experiential space-time relativity: a synopsis." Man-Environment Systems, 15(1):9-13.

Fischer, Roland. 1971.
"Time: a biocybernetic and psychopharmacological

approach." In Jiri Zeman (ed.), Time in Science and Philosophy, pp. 165-178. New York: Elsevier.

Forer, Pip. 1978.
"Time-space and area in the city of the plains." In Tommy Carlstein, Don Parkes, and Nigel Thrift (eds.), Making Sense of Time, pp. 97-118. New York: Wiley.

Forman, Frieda. 1988.
Women and Time. Elmsford, NY: Pergamon.

Forman, Frieda (ed.). 1988.
Taking Our Time: Feminist Perspectives on Temporality. Elmsford, NY: Pergamon.

Fraser, J.T. 1986.
"The problems of exporting Faust." In J.T. Fraser, N. Lawrence, and F.C. Haber (eds.), Time, Science, and Society in China and the West: The Study of Time V, pp. 1-20. Amherst, MA: University of Massachusetts Press.

Gadamer, Hans-Georg. 1966.
"Notes on planning for the future." Daedalus, 95(2):572-589.

Galtung, Johan. 1969.
"Diachronic correlation, process analysis and causal analysis: the quest for a diachronic, nomothetic social science." Quantity and Quality, 4:55-94.

Galtung, Johan. 1970.
"The social sciences: an essay on polarization and intergration." In Klaus Knorr and James N. Rosenau (eds.), Contending Approaches to International Politics, pp. 243-285. Princeton, NJ: Princeton University Press.

Gardner, Martin. 1979.
The Ambidextrous Universe. New York: Scribner.

Goldberg, M. 1964.
"Human divisions of time: a general introduction." Antioch Review, 24:7-18.

Goldblatt, Robert. 1987.
Logics of Time and Computation. Chicago, IL: University of Chicago Press.

Golledge, Reginald G. 1978.
"Learning about urban environments." In Tommy Carlstein, Don Parkes, and Nigel Thrift (eds.), Making Sense of Time, pp. 76-98. New York: Wiley.

Gonseth, Ferdinand. 1971.
"From the measurement of time to the method of research." In Jiri Zeman (ed.), Time in Science and Philosophy, pp. 277-305.

Gooddy, William. 1988.
Time and the Nervous System. New York: Praeger.

Gras, Alain. 1979.
"Time of evolution and the spirit of the times." Diogenes, 108 (Winter):57-83.

Gregory, Derek. 1978.
"Social change and spatial structures." In Tommy Carlstein, Don Parkes, and Nigel Thrift (eds.), Making Sense of Time, pp. 38-46. New York: Wiley.

Gribbin, John. 1979.
Timewarps. New York: Delacorte Press/Eleanor Friede.

Gribbin, John. 1981.
Future Worlds. New York: Plenum.

Guantao, Jin, Fan Dainian, Fan Hongye, and Liu Qingfeng. 1986.
"The evolution of Chinese science and technology." In J.T. Fraser, N. Lawrence, and F.C. Haber (eds.), Time, Science, and Society in China and the West: The Study of Time V, pp. 170-180. Amherst, MA: University of Massachusetts Press.

Haber, Francis C. 1986.
"Time, technology, religion, and productivity values in early modern Europe." In J.T. Fraser, N. Lawerence, and F.C. Haber (eds.), Time, Science, and Society in China and the West: The Study of Time V, pp. 79-92. Amherst MA: University of Massachusetts Press.

Johnson, M.C. 1952.
Time and the Universe for the Scientific Conscience. Cambridge, UK: Cambridge University Press.

Kaufman, Herbert. 1985.
Time, Chance, and Organizations: Natural Selection in a Perilous Environment. Chatham, NJ: Chatham House.

Kelly, Janice R., and Joseph E. McGrath. 1988.
On Time and Method. Newbury Park, CA: Sage Publications.

Landes, David S. 1983.
Revolution in Time: Clocks and the Making of the Modern World. Cambridge, MA: Harvard University Press.

Lawrence, Nathaniel. 1975.
"Temporal passage and spatial metaphor." In J.T. Fraser and N. Lawrence (eds.), The Study of Time II, pp. 196-205. New York: Springer-Verlag.

Lawrence, Nathaniel. 1978.
"Levels of language in discourse about time." In J.T. Fraser, N. Lawrence, and D. Park (eds.), The Study of Time III, pp. 22-49. New York: Springer-Verlag.

Lowenthal, David. 1985.
The Past is a Foreign Country. Cambridge, UK: Cambridge University Press.

Lynch, Kevin. 1972.
What Time is This Place? Cambridge, MA: MIT Press.

Macey, Samuel L. 1978.
"The changing iconography of father time." In J.T. Fraser, N. Lawrence, and D. Park (eds.), The Study of Time III, pp. 540-575. New York: Springer-Verlag.

Macey, Samuel L. 1980.
Clocks and the Cosmos: Time in Western Life and Thought. Hamden, CT: Archon Books.

Macey, Samuel L. 1987.
Patriarchs of Time: Dualism in Saturn-Cronus, Father Time, the Watchmaker God, and Father Christmas. Athens, GA: University of Georgia Press.

Macvey, John. 1988.
Time Travel: The Prospect of Travelling Through Time. Chelsea, MI: Scarborough House.

Marder, L. 1971.
Time and the Space-Traveller. London, UK: Allen and Unwin.

Melbin, Murray. 1978.
"City rhythms." In J.T. Fraser, N. Lawrence, and D. Park (eds.), The Study of Time III, pp. 444-465. New York: Springer-Verlag.

Michon, John A. 1986.
"J.T. Fraser's 'levels of temporality' as cognitive representations." In J.T. Fraser, N. Lawrence, and F.C. Haber (eds.), Time, Science, and Society in China and the West: The Study of Time V, pp. 51-66. Amherst, MA: University of Massachusetts Press.

Nystrom, Christine L. 1977.
"Immediate man: they symbolic environment of fanaticism." Etc., 34(1):19-34.

Ott, H. Walter. 1981.
"The role of the fine arts in the development of time consciousness." Educational Theory, 31:125-136.

Panikkar, Raimundo. 1978.
"Time and sacrifice: the sacrifice of time and the ritual of modernity." In J.T. Fraser, N. Lawrence, and D. Park (eds.), The Study of Time III, pp. 683-725. New York: Springer-Verlag.

Parkes, Don, and Nigel Thrift. 1978.
"Putting time in its place." In Tommy Carlstein, Don Parkes, and Nigel Thrift (eds.), Making Sense of Time, pp. 119-129. New York: Wiley.

Parr, J.B. 1981.
"Temporal change in a central-place system." Environment and Planning A, 13:97-118.

Pawley, Martin. 1974.
The Private Future: Causes and Consequences of Community Collapse in the West. New York: Random House.

Prince, Hugh. 1978.
"Time and historical geography." In Tommy Carlstein, Don Parkes, and Nigel Thrift (eds.), Making Sense of Time, pp. 17-37. New York: Wiley.

Renzong, Qiu. 1986.
"Cultural and intellectual attitudes that prevented the spontaneous emergence of modern science in China." In J.T. Fraser, N. Lawrence, and F.C. Haber (eds.), Time, Science, and Society in China and the West: The Study of Time V, pp. 181-184. Amherst MA: University of Massachusetts Press.

Rice, Paul. 1989.
Time Source. Berkeley, CA: Ten Speed Press.

Rose, Kenneth Jon. 1988.
The Body in Time. New York: Wiley.

Rosenberg, G.D., et al. (eds.). 1975.
Growth Rhythms and the History of the Earth's Rotation. New York: Wiley.

Shackle, G.L.S. 1978.
"Time, choice and uncertainty." In Tommy Carlstein, Don Parkes, and Nigel Thrift (eds.) Making Sense of Time, pp. 47-55. New York: Wiley.

Shalom, Albert. 1985.
The Body/Mind Conceptual Framework and the Problem of Personal Identity. Atlantic Highlands, NJ: Humanities.

Sperns, D.T., and A. Weinberg. 1980.
"Time and decentralization." Daedalus, Winter.

Sturm, Herta, and Marianne Grewe-Partsch. 1980.
"Time, television, and women's career perspectives." Communication Research & Broadcasting, No. 3, pp. 85-92.

Tuan, Yi-Fu. 1978.
"Space, time, place: a humanistic frame." Im Tommy Carlstein, Don Parkes, and Nigel Thrift (eds.), Making Sense of Time, pp. 9-16. New York: Wiley.

Ungvari, T. 1972.
"Time and the moderm self: a change in dramatic form." In J.T. Fraser, F.C. Haber, and G.H. Muller (eds.), The Study of Time, pp. 470-478. New York: Springer-Verlag.

Vinsrygg, Synnove. 1986.
"Time in archaelogical thought: China and the West." In J.T. Fraser, N. Lawrence, and F.C. Haber (eds.), Time,

Science, and Society in China and the West -- The Study of Time V, pp. 225-240. Amherst, MA: University of Massachusetts Press.

Warner, Rebecca. 1988.
"Rhythm in social interaction." In Joseph E. McGrath (ed.), The Social Psychology of Time: New Perspectives, pp. 63-88. Newbury Park, CA: Sage.

Wood, David C. 1989.
The Deconstruction of Time. Atlantic Highlands, NJ: Humanities.

Author Index

About the Author

T.K. DAS is Associate Professor of Strategic Management in the School of Business and Public Administration, Baruch College, City University of New York, and concurrently a member of the University's Doctoral Faculty. He has also taught at Texas Tech University and California State University at Los Angeles.

Professor Das has a Ph.D. in Management from the Graduate School of Management, University of California at Los Angeles (UCLA). He also holds two master's degrees in Management and Mathematics and a bachelor's degree in Physics.

He has published several books and monographs in management, including *The Subjective Side of Strategy Making: Future Orientations and Perceptions of Executives* (Praeger, New York, 1986) and, as co-editor, *Human Resource Management and Productivity: State of the Art and Future Prospects*, Vols. I and II (1984, 1985). He is also the author of over 90 articles in academic and professional journals.

Dr. Das has had extensive experience as a business executive. In recent years he has made presentations of his scholarly research and business insights to both academic and practitioner audiences, and has chaired sessions in various conferences, both in the U.S.A. and abroad.

Among Dr. Das's professional affiliations are: International Society for the Study of Time, The Project on Time and Organization, World Future Society, Planning Forum, Strategic Management Society, Academy of Management, The Institute of Management Sciences, and The Society of Management Science and Applied Cybernetics.